KT-152-397

PROSTATE CANCER

BRITISH MEDICAL ASSOCIATION

1003609

PROSTATE CANCER
A Case Report

GILLES PLOURDE
Department of Pharmacology and Physiology
Faculty of Medicine
University of Montreal, Montreal, QC, Canada

WITHDRAWN FROM LIBRARY
BRITISH MEDICAL ASSOCIATION

ELSEVIER

ACADEMIC PRESS
An imprint of Elsevier

Academic Press is an imprint of Elsevier
125 London Wall, London EC2Y 5AS, United Kingdom
525 B Street, Suite 1650, San Diego, CA 92101, United States
50 Hampshire Street, 5th Floor, Cambridge, MA 02139, United States
The Boulevard, Langford Lane, Kidlington, Oxford OX5 1GB, United Kingdom

© 2018 Elsevier Inc. All rights reserved.

No part of this publication may be reproduced or transmitted in any form or by any means, electronic
or mechanical, including photocopying, recording, or any information storage and retrieval system,
without permission in writing from the publisher. Details on how to seek permission, further
information about the Publisher's permissions policies and our arrangements with organizations such
as the Copyright Clearance Center and the Copyright Licensing Agency, can be found at our website:
www.elsevier.com/permissions.

This book and the individual contributions contained in it are protected under copyright by the
Publisher (other than as may be noted herein).

Notices
Knowledge and best practice in this field are constantly changing. As new research and experience
broaden our understanding, changes in research methods, professional practices, or medical treatment
may become necessary.

Practitioners and researchers must always rely on their own experience and knowledge in evaluating
and using any information, methods, compounds, or experiments described herein. In using such
information or methods they should be mindful of their own safety and the safety of others, including
parties for whom they have a professional responsibility.

To the fullest extent of the law, neither the Publisher nor the authors, contributors, or editors, assume
any liability for any injury and/or damage to persons or property as a matter of products liability,
negligence or otherwise, or from any use or operation of any methods, products, instructions, or ideas
contained in the material herein.

Library of Congress Cataloging-in-Publication Data
A catalog record for this book is available from the Library of Congress

British Library Cataloguing-in-Publication Data
A catalogue record for this book is available from the British Library

ISBN: 978-0-12-815966-8

For information on all Academic Press publications
visit our website at https://www.elsevier.com/books-and-journals

Working together
to grow libraries in
developing countries

www.elsevier.com • www.bookaid.org

Publisher: Stacy Masucci
Acquisition Editor: Rafael Teixeira
Editorial Project Manager: Tracy Tufaga
Production Project Manager: Punithavathy Govindaradjane
Cover Designer: Christian Bilbow

Typeset by SPi Global, India

CONTENTS

ABOUT THE AUTHOR

Dr. Plourde is a senior clinic evaluator for Health Canada's regulatory agency. He is an associate professor in the Department of Clinical Pharmacology and Physiology, Faculty of Medicine, at the University of Montreal and at the School of Physical Activity Sciences, Faculty of Health Sciences, at the University of Ottawa. He holds a doctorate in medicine (MD) from the University of Montreal and a PhD in experimental medicine from Laval University. His studies led him to publish several articles in highly rated, peer-reviewed journals and to make numerous presentations for prestigious associations at both national and international levels. He has more than 12 years' experience working on drug safety, especially for biologic products involved in the treatment of various cancers. For more than 4 years, he has suffered from prostate cancer with metastasis to the lungs and bones. Therefore, during the past 4 years, he has learned a lot about prostate cancer from the medical literature and discussions with various specialists. The case report presented in this book represents its own medical history. This book will provide the most relevant and up-to-date information on the various aspects involved in the screening, diagnosis, and treatment of prostate cancer.

FOREWORD

The word *cancer* is not something a patient wants to hear at the doctor's office. This word is associated with anxiety for both the patient and the people around him. The patient will often feel helpless in the face of this ordeal and might not always understand the choices that are available. *What do I have to do? What are the alternatives for me? What are my chances of survival? Will I become a burden for my family? For society?*

Although all these questions are legitimate, we must remember that in many cases, early detection significantly increases chances of patient survival. Nevertheless, patients often delay taking the necessary tests, thereby reducing their chances of recovery. The traditional roles of men in society have made them reluctant to meet with their physicians to talk about health issues. Men often try to convince themselves that the pain or incapacity is transient and that everything will return to normal soon. Therefore the diagnosis is postponed, and the effectiveness of a possible intervention is reduced. This is especially true when the pain is located in the region of the human reproductive tract.

Knowing that one out of every five new cancer diagnoses involves the prostate, it is important that men be vigilant and consult a physician as soon as possible at the onset of symptoms. However, sometimes the diagnosis is made too late, and the cancer has already settled with the presence of metastases. The presence of metastases greatly reduces the prognosis, so in order to keep the patient in the best possible condition, different choices of treatment are made available to him.

To make the best decisions possible, it is important for patients to know the different options available, along with their benefits and limitations. Using real-life examples, Dr. Plourde presents the different possibilities available to patients according to the degree of the progression of the disease and allows readers to think about the different possibilities. In this way, the patient will better understand the ins and outs of his situation and will be able to make informed choices for following these treatments.

In addition, the author also suggests resources where patients can find additional information. By explaining the different alternatives according to recent data in the literature, the author allows patients and doctors to enter into a dialogue to optimize treatment according to the chances of survival and, above all, the patient's convictions. Together, the patient and physician will develop a partnership in the treatment of the illness.

This book is therefore a call for cooperation among the different parties so that the prescribed treatment is most effective while also becoming aware of the consequences of these choices.

Guy Rousseau

INTRODUCTION

The prostate is the part of the male reproductive system that produces seminal fluid, which is one of the components of sperm. Prostate cells can be altered in the presence of various factors and can evolve into malignancy or cancer. Currently, one man in five may be diagnosed with prostate cancer during his lifetime [1–4]. Although the 5-year survival rate is nearly 100% for stages I to III, it drops to 29% for metastatic stage IV cancers. Prostate cancer represents 24% of new cancer cases among men, placing it as the most frequent cancer diagnosed in men after skin cancer [4]. For most cancers, many factors influence the cell susceptibility to turn or become dysfunctional or abnormal, leading to the onset of the disease. In this book, readers will learn about the prevention, screening, treatment, and follow-up procedures associated with prostate cancer, as well as the complications associated the disease. This information will be provided via the presentation of real case reports with questions and answers from a patient diagnosed with invasive prostate cancer and its associated complications.

The signs and symptoms of prostate cancer are often linked to difficulty with urinary functions, including the frequency and urgency (i.e., pressing desire) of urination, such as an inability to perform a complete emptying of the bladder (see Case Report #1). The presence of blood in urine or sperm, painful ejaculations, and abdominal or pelvic pains are also warning signs and symptoms of prostate cancer [5] as described in the case of our 52-year-old patient presented here. As the disease develops the above signs and symptoms become more important; and other symptoms including weight loss, tiredness, anemia, loss of bladder control, and even bone pain also occurred in our patient. With these signs and symptoms, we are more and more confident that our patient is presenting with prostate cancer and while making the main diagnosis, we should always consider the differential diagnosis that is discussed in the Case Report #2.

The main early detection and confirmation methods of prostate cancer are the digital rectal examination (DRE), the prostatic specific antigen (PSA), and the transrectal ultrasound with the fine needle prostate biopsies (see Case Report #1). Recently, new biomarkers have become available, and the information about their most recent developments as relevant to the screening, diagnosis, follow-up of disease progression, and response to treatment will be discussed in Case Report #9 and others, as applicable.

Blood tests and imaging techniques are used to determine the staging of prostate cancer; that is, whether the cancer is located in the prostate only or has expanded to other organs such as the lungs, liver, bones, and lymph nodes when the diagnosis is confirmed. Biopsies are used to describe the organization of the prostate cancer (i.e., whether or not it is differentiated; see Case Report #1).

The adenocarcinomas account for 95% of prostate cancers. This type of cancer usually changes slowly, allowing for a quick treatment in the majority of cases in order to limit the adverse effects of the disease. In general the most currently used treatments for this type of cancer are surgery, radiation therapy, hormone therapy, and of course, chemotherapy (see Case Report #3). Despite better awareness, prevention, and treatments, a 28% increase in prostate cancer cases is predicted for 2028–32 [4].

As will be discussed in Case Report #4, there are two categories of risk factors: those that can be avoided and those that cannot. For example, in prostate cancer, factors such as age, ethnicity, genetic predispositions, and family history cannot be modified, so they are considered as internal factors [1–4].

Despite the lack of a strong causal link for the modifiable risk factors, it seems that a diet rich in flaxseed and lycopene (an antioxidant) in addition to regular physical activity would contribute to the prevention of this type of cancer [1–3]. Exposure to environmental agents such as X-rays, gamma X-rays, cadmium, and arsenic also increases the risk of prostate cancer [1–3]. This year, among the 41,000 deaths in men due to cancer, approximately 10% will be related to prostate cancer. In Case Report #4, we will discuss the main external risk factors involved in prostate cancer and the precautions that should be taken to prevent them.

The main objective of this book is to discuss strategies for the prevention and treatment of prostate cancer, as well as the adverse effects of the medication used to treat this serious health problem in order to provide patients and various partners (e.g., health care providers, nurses, medical staff, family, and others) with the tools to better prevent and manage this condition. The goal is also to present the most appropriate information that will allow patients and their families to make informed decisions, as well as to perform a benefit/risk analysis regarding the choice of their treatments. We will also discuss the paraneoplastic syndromes that are often associated with cancer. In particular, during his treatment our patient developed persistent vertigo and headaches that will be discussed in Case Report #5. In Case Report #6, we will discuss the main problem associated with his invasive prostate cancer, which is bone metastasis, and we will provide information on treatment modalities of bone metastasis.

Another important objective of this book is to provide relevant information on diet, physical activities, and stress management in order to improve the quality of life of patients suffering from prostate cancer (see Case Report #7). Although the information is presented in the form of real case reports of a patient suffering from prostate cancer, it can also be applied to patients suffering from other forms of cancer.

Medicine is still a very limited science, and the treatment of cancer is increasingly complex. This book illustrates the importance of a better use of science in the screening, diagnosis, treatment, and follow-up of patients with prostate cancer. Regarding this, we will discuss in Case Report #8 the most promising treatment modalities for prostate cancer including: immunotherapy (checkpoint inhibitors), gene therapy, nanotechnology, vaccines, and oncolytic virotherapy. We also discuss the use of biomarkers that may better facilitate cancer screening by identifying the severity of the disease at the time of diagnosis and easing the assessment of the response to treatment (see Case Report #9).

Obviously, we cannot conclude this book without discussing the pros and cons of euthanasia and assisted suicide and the concerns raised by both patients and society as a whole on this issue (see Case Report #10). Similarly, we will discuss the role of palliative care and palliative sedation when other treatment modalities fail to control the progression of cancer and when health care providers are no longer able to manage the physical and psychological pain the patients are experiencing at the end of their lives.

This book will conclude with a discussion of cannabis, which has long been used for medicinal purposes. We will discuss the favorable outcomes associated with cannabis that include: the beneficial effects on chemotherapy-induced nausea and vomiting, as well as cancer-related psychological and physical pain. The benefits of cannabis in the treatment of anorexia, insomnia, and anxiety are also discussed. Finally, cannabinoids have shown antineoplastic effects in preclinical studies in a wide range of cancer cells and in some animal models. This will also be discussed in the Case Report #10.

Of course, compared to the last books by the same author, he will draw up a list of resources for patients and health care professionals wishing to deepen their knowledge of the topics covered in the following case reports. By combining in a single volume the latest research as well as data from the different national and international clinical practice guidelines concerning the prevention and treatment of prostate cancer the author hopes to establish the basis of a health program to improve the prevention, treatment, and quality of life of patients with prostate cancer. The author wishes to encourage the cooperation of all partners involved in prevention and treatment activities, including universities, regulatory agencies, governments, health care professionals, national and international partners, patients, patient representatives, and the general population in the fight against prostate cancer.

REFERENCES

[1] Canadian Cancer Society. What is prostate cancer? Available at: http://www.cancer.ca/en/cancer-information/cancer-type/prostate/prostate-cancer/?region=on; 2017.

[2] American Cancer Society. Key statistics for prostate cancer. Available at: https://www.cancer.org/cancer/prostate-cancer/about/key-statistics.html; 2017.

[3] Canadian Cancer Society's Advisory Committee on Cancer Statistics. 2017. Available at: http://www.cancer.ca/en/cancer-information/cancer-101/canadian-cancer-statistics-publication/?region=bc.

[4] Public Health Agency of Canada. Health promotion and prevention of chronic disease in Canada. Res Policy Pract 2015;35(Suppl. 1):200.

[5] Public Health Agency of Canada. What should I know about prostate cancer? http://www.phac-aspc.gc.ca/cd-mc/cancer/prostate_cancer_about-cancer_prostate_sujet-eng.php; 2017.

CHAPTER 1

Case Report #1—Presentation of the Patient

INTRODUCTION

A 52-year-old man with a 5-month history of urinary symptoms (e.g., difficulty initiating urination, reduced urinary flow, incomplete emptying of the bladder, frequent urination especially at night [prostatism]) visited a walk in clinic at the end of March 2014 due to the onset of pelvic, abdominal, lower back (lumbar) and hip pain. The health care provider performed a digital rectal examination (DRE) and found that the volume and consistency of the prostate was normal. According to the description made by the health care provider the prostate was similar to what was seen during the routine physical examination performed 5 years ago by patient's family doctor.

Question #1: Based on this preliminary information what should be the most probable diagnostic?

(A) **Chronic prostatitis**
(B) **Benign prostate hyperplasia (BPH)**
(C) **Prostate cancer**

Answer: C.

(A) **Chronic prostatitis.** Prostatitis represents the swelling and inflammation of the prostate gland, a walnut-sized gland located directly below the bladder in men. Prostatitis often causes painful or difficult urination. Other symptoms include pain in the groin, pelvic area, or genitals, as well as flu-like symptoms in some cases. Prostatitis affects men of all ages but tends to be more common in men aged 50 or younger. This condition can be due to many causes, which often are unidentified. If prostatitis is caused by a bacterial infection, it can usually be treated with antibiotics. Depending on the cause, prostatitis can come on gradually or suddenly. It might improve quickly, either on its own or with treatment. Some types of prostatitis last for months or keep recurring (chronic prostatitis).

Prostate Cancer
https://doi.org/10.1016/B978-0-12-815966-8.00001-1

© 2018 Elsevier Inc.
All rights reserved.

For our patient a diagnosis of chronic prostatitis was suggested considering the duration of symptoms. The patient was prescribed Levaquin 750 mg per day for 1 month to treat this suggested chronic prostatitis and Rapaflo 8 mg per day to allow for a relief of urinary symptoms. The health care provider told the patient to take nonsteroidal antiinflammatory drugs (NSAIDs) to relieve his pain.

A prostate-specific antigen (PSA) level and a urinary analysis and culture were requested. The PSA level came back to 25.6 ng/mL; normal is between 0 and 4 ng/mL for a patient this age. At this level the PSA was already strongly evocative of prostate cancer. Due to the symptoms and high PSA level the health care provider made a consultation with an urologist.

On June 9, 2014 a urologist performed a prostate physical examination and surprisingly noted that the prostate was enlarged mainly on the right side. Then he decided to prescribe another month of antibiotics (i.e., Ciprofloxacin 500 mg twice a day), claiming that this antibiotic is better at treating prostatitis than the previous one. He also prescribed a continuation of the NSAIDs for the pain. Obviously it was very unlikely that he dealt with a case of chronic prostatitis, nevertheless, this urologist did nothing immediately to eliminate the possibility of prostate cancer being more concerned about a chronic prostatitis. For more information on the use of PSA as a screening and management tools for prostate cancer, please see Item C and consult Case Report #9 on biomarkers.

(B) Benign prostate hyperplasia (BPH). Also called benign prostatic hypertrophy. BPH is an enlargement of the prostate gland due to an increased number of cells (hyperplasia). Most of the growth occurs in the transition zone of the prostate. The prostate naturally gets larger as men age; in fact, almost all men will have some prostate enlargement by the age of 70. Other than increasing age, there are no risk factors for BPH. For now, researchers cannot confirm whether BPH increases the risk of prostate cancer. With the rapid progression in PSA levels (see C) in such a limited period of time, it is highly unlikely that the patient had BPH.

(C) Prostate cancer. Prostate cancer is a malignant tumor that starts in the cells of the prostate. Malignant means that it can spread, or metastasize, to other parts of the body. Prostate cancer is the second-most common cancer in Canadian men [1,2]. It usually grows slowly and can often be completely removed or managed successfully. Cells in the prostate sometimes change and no longer grow or behave normally. These changes may lead to noncancerous or benign conditions such as prostatitis and BPH [1,2].

Changes to the cells of the prostate can also cause precancerous conditions. This means that while the cells do not yet indicate cancer, there is a higher chance these abnormal changes will become cancerous.

The precancerous conditions that can develop in the prostate are prostatic intraepithelial neoplasia (PIN), proliferative inflammatory atrophy (PIA), and atypical small acinar proliferation (ASAP). Most often, prostate cancer starts in the glandular cells of the prostate. This type of cancer is called adenocarcinoma of the prostate. Rare types of prostate cancer can also develop; these include transitional cell carcinoma and sarcoma. In mid-April 2014, another blood test was requested and came back normal for the patient's hemoglobin, white blood cells, and platelets, as well as for the liver and urinary functions. However, the PSA had jumped to 79.3, which was increasingly evocative of prostate cancer, in which case the PSA continues to rise very quickly.

Comment #1: For the reader's information the PSA is an enzyme, a protease, and its gene is dependent on androgens [3]. The rise in levels observed in prostate cancer are likely due to an increase in the passage of the antigen through the basal membrane, which is broken by the cancer. This protease is produced, in theory, exclusively by the prostate; however, special situations have been described in the course of which the PSA concentration is increased [3]. For example, there is a rise in the PSA in about a third of breast cancer diagnoses and its concentration can also be increased significantly in the case of hepatitis, though not at the level observed in our patients.

Comment #2: About 70% of total serum PSA that circulates is bound to the blood proteins form, and 30% is in free form. The free form increases in the case of BPH, while the bound form increases in the case of prostate cancer [3]. The PSA free/total ratio changes greatly in the case of prostate cancer (see Case Report #9). For men under 70 a PSA level <3 ng/mL is considered to be normal. For men over 70 the PSA levels rise slightly with age so that a value of 6.5 ng/mL can be considered reassuring. The PSA alone does not allow the exact diagnosis of prostate cancer, but it is certainly one of the most important risk factors for prostate cancer according to Mondo et al. [3]. For example the PSA can be high in various prostate pathologies including prostate adenoma and prostatitis (prostate inflammation/infection). It can be elevated after a rectal examination (but not in a significant way), cystoscopy, establishment of a house probe, or intervention on the prostate.

Comment #3: For the benefits of the reader, PSA is the most widely used biomarker for the early detection of prostate cancer. Since the introduction of PSA testing, prostate cancer diagnoses have increased, but at the same time the number of patients dying from the disease has decreased [3]. Furthermore, higher PSA levels are associated with a higher risk of cancer, high-grade disease, high tumor stage, and the presence of metastatic disease as in our patient [3].

However, according to the medical literature PSA, does not represent an ideal biomarker [3]. First, commercial assays measuring PSA are not standardized for direct comparison, so repeat testing is usually necessary. Second, PSA levels are not specific to prostate cancer and can be modulated by many factors, such as age, infection, trauma, ejaculation, instrumentation, and medication use (e.g., 5-alpha-reductase inhibitors and corticosteroids). Third, there is no absolute value below which there is a negligible risk of prostate cancer, and PSA levels cannot distinguish between indolent and aggressive diseases. In the prostate cancer prevention trial, approximately 15% of men with a PSA below 4 ng/mL were at risk for prostate cancer, while 15% of these men had a high-grade disease [3].

However, when the PSA level was <1 ng/mL, the risk of high-grade disease is very low. Moreover, PSA levels above the traditional cutoff of 4–10 ng/mL reveal the presence of cancer on biopsies in only 25%–30% of patients. Hence there is no PSA cutoff point with high sensitivity and specificity for prostate cancer monitoring in healthy men, but there is rather a continuum of prostate risk at all values of PSA [3]. However, a rapid and continuous increase in PSA levels (short doubling time) as seen with our patient is always alarming.

Prostate cancer screening with PSA levels has been a subject of debate and controversy due to its potential toward overdetection and overtreatment, which can induce patient anxiety. Indeed the ability of PSA levels to reduce mortality has produced mixed results in recent randomized screening trials. Uncertainty also exists in the practical considerations of testing, such as the age at which to initiate and discontinue the testing, along with its frequency [3].

Various guidelines addressing prostate screening have highlighted these issues of uncertainty, prompting the US Preventive Services Task Force to recommend *against* the use of PSA levels for screening in 2012 [4]. Nonetheless, PSA levels still remain the first-line biomarker option for the detection of prostate cancer. In a recent review of the Canadian Task Force on Preventive Health Care, for which our patient act as an external reviewer has recommended: not screening with PSA in patients of <55 years-old and patients older than 70 years old. This suggests that the PSA is not recommended as a screening tool in patients without symptoms or prostate disease [5].

On the other hand the higher the PSA, the higher the possibility of a cancer expansion beyond the prostate. Therefore it is important to remember that clinical examination remains irreplaceable in the case of this disease. In order to increase the sensitivity of this test, it is suggested to use the PSA free/total ratio [3]. Thus for a PSA ranging between 4 and 10 ng/mL and a ratio <5% the probability of a prostate cancer is strongly suspected. Conversely a ratio >30% is more in favor of a benign prostate disease such as benign prostate hypertrophy [3]. Although this ratio has not been measured in our patient the PSA value of 79.3 ng/mL was already very suggestive of prostate cancer. Therefore C is the best answer.

Question #2: As mentioned earlier the symptoms did not improve with the second round of antibiotics, which goes against the diagnosis of prostatitis. Now we have a patient with high PSA level, bone pains, and a worsening of urinary symptoms to support an initial diagnosis of prostate cancer.

What other tests should be done to further support our initial diagnosis?

(A) Transrectal ultrasound
(B) Cystoscopy
(C) Abdominal and pelvic ultrasounds
(D) Prostate biopsy
(E) Magnetic resonance imaging or axial CT scan of the whole body
(F) Bone scintigraphy or bone scan
(G) All of the above

Answer: G.

(A) Transrectal ultrasound. A transrectal ultrasound may also be called prostate sonogram or endorectal ultrasound. It is used to look at the prostate and tissues around it. An ultrasound transducer (also called a probe) sends sound waves through the wall of the rectum and into the prostate and surrounding tissue. A computer analyzes the wave patterns (echoes) as they bounce off the organs and convert them into an image that health care providers view on a video screen. As a general rule, any abnormal PSA values should be supported by a transrectal ultrasound, which allows for searching for images evoking the presence of cancerous tissue. However, this test was carried out a few months later in combination with prostate biopsies, which are the gold standard for making a diagnosis (see below).

(B) Cystoscopy. A cystoscopy is a procedure that allows health care providers to examine, take samples from, or treat problems in the bladder and urethra. The health care providers use an endoscope, which is a thin tube-like instrument with a light, lens, and camera used to examine organs or structure in the body. The cystoscope can be straight and stiff (rigid cystoscope) or it can bend (flexible cystoscope). Health care providers most often use a flexible cystoscope.

On June 9, 2014 a cystoscopy was performed in our patient; it showed dilated veins inside the bladder, but no active bleeding or mass. This procedure was very difficult and caused heavy bleeding to the point that our patient became highly anemic. The leading hypothesis to explain this difficult exam in cystoscopy would be secondary to the urinary tract compression by enlarged lymph nodes.

(C) Abdominal and pelvic ultrasounds. These tests were performed on June 26, 2014, which confirms the presence of many enlarged lymph nodes in the abdominal and pelvic areas, as well as an enlarged and severe right kidney hydronephrosis (filled fluid kidney). This hydronephrosis would likely be due to a blockage caused by enlarged lymph nodes close to the bladder and the prostate. Nevertheless, this hydronephrosis was not reassuring for the patient, as it makes important perturbations in its urinary functions and was an important source of anxiety putting his right kidney at risk of failure.

(D) Prostate biopsy. A prostate biopsy is used to determine whether any suspicious-looking tissues are cancerous. A biopsy is conducted when an abnormal lump is found during a digital rectal examination (DRE) or if the PSA blood test reveals high levels of PSA, as in our patient. The prostate biopsy was performed on July 23, 2014, along with the transrectal ultrasound. The Gleason index (i.e., an index commonly used to study the prostate cancer severity, which is done by examining the appearance of prostate cells under a microscope), showed a result of 9/10 (5 + 4), thus strongly suggesting the presence of an aggressive and intrusive prostatic adenocarcinoma.

These Gleason criteria describe the appearance and organization of cells; thus 5 indicates that cells have a very different appearance from normal cells, while 4 means complete cell disorganization and therefore a very advanced cancer. High-grade cancer can be very difficult to treat and can reappear quickly. There are other forms of nonaggressive prostate cancer that do not usually cause death. Unfortunately, our patient has an aggressive form of cancer associated with a very bad prognosis. Also note that this patient had shown symptoms of tiredness, a low-grade fever, and a weight loss of >15 pounds in 6 months.

(E) Magnetic resonance imaging or axial CT scan of the whole body. In the presence of invasive prostate cancer, it is essential to assess for the cancer extension in order to determine the therapeutic approach. One of the above imaging tests is performed in order to evaluate the possible ganglionic or metastatic extension to other organs (e.g., lungs, liver, and bones) that are the main sites for metastasis in prostate cancer. This test is also very useful to determine the grade of cancer.

The whole body CT scan performed on July 2, 2014, confirmed the presence of many enlarged lymph nodes in the body, in addition to a severe increase in the volume of the right kidney (i.e., right nephromegaly and a severe right kidney hydronephrosis). The CT scan also showed many lung nodules that the radiologist described as being too many to be listed but were very suggestive of multiple lung metastases; however, this radiological test does not identify liver nodules. These results clearly demonstrated a very invasive prostate cancer that invaded the lungs, lymph nodes, and bones but not the liver.

Comment #1: Staging. It is an effective way to classify the cancer evolution based on its origin, size, degree of infiltration to the surrounding tissue, and ability to spread to the lymph nodes and other parts of the body. In general, four stages exist (i.e., I–IV) to describe different types of cancer. Stage I is the early stage where cells are located but have not spread to lymph nodes or other anatomic organs. Stage II is also localized cancer in the organ of origin, but it may extend to the surrounding tissue. Stage III indicates the cancer has spread deeper and is present at the lymph node levels. Finally, when the cancer is classified as stage IV, it has become metastatic, which is more commonly known as generalized cancer, as in our patient [6,7].

As cells, cancerous or not, can travel inside blood and lymphatic channels, they can therefore appear anywhere in the body. When one of these cells appears far from the tumors of origin and proliferates, it forms a second tumor called metastasis. The presence of metastases correlate usually positively with a wrong prognostic.

Most cases of death from cancer in humans are not due to the primary tumor, but to the metastases or subsequent tumors that grow in a different anatomic site (e.g., secondary, tertiary tumors, etc.). It is therefore important to determine the grade of cancer at diagnosis, as it will determine the choice of treatment, the aggressiveness of the treatment, and the patient prognosis [6,7].

(F) Bone scintigraphy or bone scan. A bone scan is a nuclear medicine imaging test that uses bone-seeking radioactive materials or tracers (radiopharmaceuticals) and a computer to create an image of the skeleton (bones). A bone scan looks at the bones to see if there are any abnormalities, such as a fracture, tumor, metastases, or infection.

The bone lesions are often highlighted several months before being visible on X-rays, which is why a bone scan should be performed. This assessment is conducted in patients suffering from bone pain and has a PSA level >20 ng/mL or a Gleason score >7 as in our patient, where there was a strong suspicion of bone metastases. This Gleason score can also serve as a benchmark for the follow-up of patients with bone metastases.

On September 8, 2014, our patient passed a bone scan that revealed the presence of multiple metastases at the skull level, left scapula, pelvic bones, spine (vertebrae), and ribs. It is the most sensitive test to identify bone metastases. The spine is the second most common site of bone metastases after the pelvis. Bone scintigraphy can detect bone metastases quicker than other diagnostic radiology methods and is of no danger to patients. However, small bone metastases may go unnoticed even with this approach.

Discussion: The oncologist in charge of this patient confirmed the diagnosis of stage IV metastatic prostate adenocarcinoma. He gave the patient a very short life expectancy (2–4 years) knowing that cancers at this stage are incurable (see Case Report #3). When metastases reach the skeleton (e.g., pelvis, ribs, vertebrae) as it did in our patient, a 1-year life expectancy is <40% and a 5-year life expectancy is <1%; therefore the prognosis mentioned above by the oncologist seems to be appropriate [6]. According to the Eastern Cooperative Oncology Group (ECOG) scale, which includes six values from 0 to 5 (where 0 represents a good state of health and 5 death), The patient is ECOG level 1, and therefore symptomatic (e.g., less able to perform regular physical activities, but able to travel alone and to do sedentary or light work, such as a desk job or household work).

The oncologist was reassuring by saying that the treatments (discussed below) would remove the metastases to the lungs, the lymph nodes, and bone metastases, which are partially correct as currently after more than

3 years of hormone therapy, the patient is still developing new bone metastasis identified on repetitive bone scans. However, the lung nodules have completely disappeared as have the lymph nodes.

However, there are conditions that may affect those probabilities, such as age of the patient, his general health condition, the extent of metastasis, the PSA level, and his response to treatment. Fortunately the patient responded well to hormone therapy for many years, but recently he has had to change his treatment for a second-line hormone therapy (see below). In addition, with all the measures used by the patient including an adequate diet, regular physical activity, and stress management to combat his cancer (see Case Report #7), his quality of life hopefully will be much better and his life expectancy much longer than predicted by the above statistics or by his oncologist.

REFERENCES

[1] Canadian Cancer Society. What is prostate cancer? Available at: http://www.cancer.ca/en/cancer-information/cancer-type/prostate/prostate-cancer/?region=on; 2017.

[2] Canadian Cancer Society's Advisory Committee on Cancer Statistics. 2017. Available at: http://www.cancer.ca/en/cancer-information/cancer-101/canadian-cancer-statistics-publication/?region=bc.

[3] Mondo DM, Roelh KA, Loeb S, et al. Which is the most important risk factor for prostate cancer: race, family history, or baseline PSA level? J Urol 2008;179:148.

[4] Moyer VA, U.S. Preventive Services Task Force. Screening for prostate cancer: U.S. Preventive Services Task Force recommendation statement. Ann Intern Med 2012;157:120–34.

[5] Bell N, Connor Gorber S, Shane A, et al. Canadian Task Force on Preventive Health Care. Recommendations on screening for prostate cancer with the prostate-specific antigen test (guidelines). CMAJ 2014;186:1225–35.

[6] American Cancer Society. Survival rates for prostate cancer. Available at: https://www.cancer.org/cancer/prostate-cancer/detection-diagnosis-staging/survival-rates.html; 2017.

[7] Canadian Cancer Society. Staging. Available at: http://www.cancer.ca/en/cancer-information/diagnosis-and-treatment/staging-and-grading/staging/?region=qc; 2017.

CHAPTER 2

Case Report #2—Other Differential Diagnoses

INTRODUCTION

On June 15, 2014, the same patient woke up with a new and sudden pain on the left side of his neck. He looked in the mirror and noticed a new enlarged mass in this area. The very worried patient made an emergency appointment with his family doctor, who also noted the presence of enlarged lymph nodes under the armpit, abdomen, and inguinal regions. Because of this mass, the patient made a quick appointment with an oncologist specializing in lymphoma at the McGill University Cancer Research Center located hundreds of kilometers from the patient's residence. Meanwhile, recent bloodwork results revealed that the PSA had risen to 173 ng/mL in the past few weeks (very short doubling time), as well as the presence of severe microcytic anemia that the patient gradually corrected by taking iron supplements.

Question #1: What should be the differential diagnosis?

(A) Viral syndrome
(B) Hodgkin lymphoma
(C) Advanced prostate cancer

Answer: C.

(A) Viral syndrome. Viral syndrome is a term uses for general symptoms of a viral infection that has no clear cause. Signs and symptoms may start slowly or suddenly and last hours to days. They can be mild to severe and can change over days or hours. The serology for the viral syndrome that was performed in our patient showed a negative HIV test, negative Epstein Barr virus (EBV), and negative cytomegalovirus (CMV) tests. Other serological tests were also negative, as well as the urinary culture. Therefore based on the presentation (see Case Report #1) and the above information, it was very unlikely to be a viral syndrome.

Prostate Cancer
https://doi.org/10.1016/B978-0-12-815966-8.00002-3

© 2018 Elsevier Inc.
All rights reserved.

(B) Hodgkin lymphoma. (HL) is a cancer that starts in the lymphocytes, which are the white blood cells of the lymphatic system. The lymphatic system works with other parts of the immune system to help the body fight infection and disease. The lymphatic system is made up of a network of lymph vessels, lymph nodes, and lymphatic organs. Lymph vessels carry lymph fluid, which contains lymphocytes and other white blood cells, antibodies, and nutrients. Lymph nodes sit along the lymph vessels and filter lymph fluid. The lymphatic organs include the spleen, thymus, adenoid, tonsils, and bone marrow. Lymphocytes develop in the bone marrow from basic cells, which are called stem cells. Stem cells develop into different types of cells that have different jobs. The main types of lymphocytes are: (1) B cells that make antibodies to fight bacteria, viruses, and other foreign materials such as fungi; (2) T cells that fight infection, destroy abnormal cells, and control the immune response; and (3) Natural killer (NK) cells that attack abnormal or foreign cells.

Lymphocytes sometimes change and can no longer grow or behave normally. These abnormal cells can form tumors called lymphomas. Hodgkin lymphoma usually starts in abnormal B cells called Hodgkin and Reed-Sternberg cells, or HRS cells. These cells are much larger than normal lymphocytes and have either a large nucleus or more than one nucleus.

Hodgkin lymphomas are divided into two main forms based on the presence of HRS cells. Classic HL means that the HRS cells are present. The classic forms of HL include nodular sclerosis HL, mixed cellularity HL, lymphocyte-rich classic HL, and lymphocyte-depleted HL. Nodular lymphocyte-predominant HL means there are very few or no HRS cells present. Because lymphocytes are found throughout the lymphatic system, HL can start almost anywhere in the body. It usually starts in a group of lymph nodes in one part of the body, most often in the chest, in the neck, or under the arms. It usually spreads in a predictable orderly way from one group of lymph nodes to the next. Eventually, it can spread to almost any tissue or organ in the body through the lymphatic system or the bloodstream.

Other cancers of the lymphatic system are called non-Hodgkin lymphoma (NHL). The HRS cells of Hodgkin lymphoma look and behave differently from non-Hodgkin lymphoma cells. Hodgkin lymphomas and non-Hodgkin lymphomas are treated differently, but such a discussion is outside the scope of this case report.

The whole body CT scan performed on our patient on July 2, 2014, confirms the presence of many enlarged lymph nodes in the body in addition to a severe increase in the volume of the right kidney (i.e., an enlarged right kidney and a severe right kidney hydronephrosis). As discussed earlier, the CT scan showed many lung nodules that were suggestive of multiple lung metastases without excluding the possibility of being small lymph nodes. Therefore, a Hodgkin lymphoma cannot be excluded, though it was less likely considering the signs and symptoms presented by the patient, which are more consistent with a primary prostate cancer diagnosis with lymph node involvement.

The family doctor also scheduled an urgent biopsy of the ganglionic mass on the neck, which was carried out a few days later on July 8, 2014. The lymph node biopsies of the left neck mass showed a possible primary prostate cancer though it couldn't yet be confirmed. In addition the possibility of lymphoma couldn't be eliminated at this time given the extent of the lymphadenopathy and the absence of a final differentiation during the pathology assessment of the lymph nodes biopsies collected on the left side of the neck.

(C) Advanced Prostate Cancer. A few weeks later, on August 4, 2014, the patient met with an oncologist who specializes in lymphoma at the McGill University Cancer Research Center. Based on the biopsies results taken from the mass on the left side of the neck and radiology exams, the specialist confirmed that the patient had an aggressive and intrusive prostate adenocarcinoma of stage T3c, which meant the tumor had extended outside the prostate and its capsule. In addition the seminal vesicles are affected with a notation N3 [+N], which confirmed that several lymph nodes are located in the abdominal, pelvic, and thoracic areas, as well as M1, which confirmed the presence of metastases in other organs far away from the prostate such as the lungs and bones but not the liver. Overall, this is a stage IV cancer, which is the highest of the four stages and the most difficult prostate cancer to treat. Therefore C is the best answer.

Comment #1: Another useful tool or biomarker to establishing a prostate cancer prognosis is the nomogram, which is an instrument that associates a set of input data to a particular outcome [1]. The predictive power of a nomogram can be superior to the PSA level alone because they combine a greater number of prognostic variables specific to an individual patient. They usually incorporate information (e.g., clinical stage, PSA levels) and pathological information (e.g., Gleason score and number of positive biopsy scores [2]). For our patient, incorporating this information into a nomogram adds little values in the diagnosis of his prostate cancer. We knew that he was stage IV with a PSA level of 173 ng/mL and a Gleason score of 9, which confirmed an invasive and undifferentiated prostate cancer.

Numerous nomograms have been developed for different clinical situations, such as treatment decision making for patients eligible for active surveillance, radical prostatectomy (RP), neurovascular bundle preservation, and pelvic lymph node dissection omission during RP or radiotherapy [1]. Posttreatment nomograms also exist, providing estimates of biochemical progression-free survival (PFS) after RP or the potential success of salvage radiation therapy after RP.

However, the use of nomograms has been criticized, particularly nomograms developed in research centers that may be unable to generalize results for the patient population [1]. Nomograms may also incorporate subjective or intermediate endpoints and could be affected by changing diagnostic procedures. No nomograms were performed for our patients; with the information we have (e.g., the high PSA level, the pathology results and bone scan), we know that the diagnosis was an advanced stage IV prostate cancer. But the information from nomograms might still be useful for other patients suffering from prostate cancer.

Discussion: With the patient's diagnosis of an aggressive [stage IV] and intrusive prostate adenocarcinoma, the next important issue is optimal cancer management. The first question to ask is what the patient should do with his diagnosis. Needless to say the choice of treatment depends on the individual; indeed, what works for one patient may not necessarily work for another patient even with the same type of cancer.

This is why it is essential to evaluate the risks and benefits of each treatment option and to choose the ones with higher positive benefit/risk ratios. A decision should be made only when all the required information is available because all treatment options have their pros and cons regarding their short- and long-term benefit/risk ratios. The main elements to be considered in choosing a treatment are the patient's quality of life, his general state of health, his life expectancy in good health, and the degree of cancer cell spread (i.e., stage of cancer). It is also important to choose between what the patient fears the most: the cancer itself or its treatment? Unfortunately the question of cost/benefit must also be taken into account, as well as each individual's willingness to participate in research projects, if available. One should know that it is often possible to adhere to programs that will help patients deal with the financial cost of the medication. Therefore when you are diagnosed with an advanced form of prostate cancer, do not hesitate to pose questions to your oncologist or medical team in order to be enrolled in these programs. This is true not only for prostate cancer but for other cancers as well.

It is important to ask these questions and take the necessary time to have an open discussion, because ultimately both the patient and his health care provider will have to make the decisions. Certainly the patient can let the doctor make the decision for him if that is his choice. No matter who makes the decision the health care provider must collaborate with the patient so that he makes a clear and well-informed decision about his treatment choices.

The objective of the third case report is to provide information regarding the main treatment modalities available (e.g., chemotherapy, hormone therapy, radiation therapy, surgery, etc.) in order to help health care providers and their patients to choose the optimal treatment approach.

Of course, because several of the treatments offered are still in development and not yet approved in Canada, patients may request them via applications for special access program from countries where the treatments are approved.

Comment #1: The special access program (SAP) provides access to nonmarketed drugs for practitioners treating patients with serious or life-threatening conditions when conventional therapies have failed, are unsuitable, or are unavailable. The SAP authorizes a manufacturer to sell a drug that cannot otherwise be sold or distributed in Canada. Drugs considered for release by the SAP include pharmaceutical, biological, and radiopharmaceutical products not yet approved for sale in Canada. Therefore some oncologic drugs that are not yet authorized in Canada but are in other countries can become available throughout the SAP program. For more information on this program, please visit http://www.hc-sc.gc.ca/dhp-mps/acces/drugs-drogues/sapf1_pasf1-eng.php.

REFERENCES

[1] Gaudreau PO, Stagg J, Soulières D, Saad F. The present and future of biomarkers in prostate cancer: proteomics, genomics, and immunology advancements. Supplementary issue: biomarkers and their essential role in the development of personalised therapies (A). Biomark Cancer 2016;8(Suppl. 2):15–33.
[2] Mondo DM, Roelh KA, Loeb S, et al. Which is the most important risk factor for prostate cancer: race, family history, or baseline PSA level? J Urol 2008;179:148.

CHAPTER 3

Case Report #3—Treatment Options

INTRODUCTION

With a diagnosis of stage IV metastatic prostate adenocarcinoma the patient was given a very short life expectancy (see Case Report #1). At this stage, unfortunately we can no longer talk of a cure for a patient; instead the prolongation of life and the quality of it become the priority. Improving the patient's quality of life is an important part of treatment; therefore the oncologist and the patient should discuss treatment options and make a benefit/risk analysis for each option in order to make informed decision about the selected treatment(s).

Question #1: What is the best treatment options available for this patient?

(A) Chemotherapy
(B) Surgical therapy
(C) Radiation therapy
(D) Brachytherapy
(E) Hormone therapy
(F) Surgical castration

Answer E.

(A) Chemotherapy. Chemotherapy is the use of cytotoxic drugs to treat cancer. It is usually a systemic therapy that circulates throughout the body and destroys cancer cells, including those that might have broken away from the primary tumor. Chemotherapy has an important role in the treatment of prostate cancer in patients who become resistant to hormone therapy (e.g., hormone-refractory or castration-resistant prostate cancer). Chemotherapy drugs, doses, and schedules vary from patient to patient. The most common chemotherapy drugs used to treat hormone-refractory prostate cancer are: docetaxel (Taxotere), mitoxantrone (Novantrone), and cabazitaxel (Jevtana); see below. The most common first-line chemotherapy used is docetaxel, which helps to reduce symptoms and prolong overall survival (OS) [1,2].

Prostate Cancer
https://doi.org/10.1016/B978-0-12-815966-8.00003-5

© 2018 Elsevier Inc.
All rights reserved.

Comment #1: Our patient consulted with another specialist of the same McGill University Cancer Research Center as before (see above) who advised him to take docetaxel (chemotherapy) as a first-line therapy in combination with hormone therapy. The specialist suggested that taking this combination early, even if the patient is not yet resistant to hormone therapy, will allow the patient to increase his life expectancy for several months. The specialist adds that our patient is still in good condition and would be able to tolerate it well. However, according to the recent medical literature, there are other options to consider with a better benefit/risk ratio than early chemotherapy. For these reasons, A is not the best option.

Comment #2: Chemotherapy is probably the most widely used cancer therapy for a great variety of cancers. Given alone or more often in combination the chemotherapeutic agents are designed to slow down or completely stop the proliferation of cancer cells; thus they are particularly effective against cells with a high rate of cell division. Unfortunately, normal cells of the body with a pace of rapid division (e.g., the intestinal cells, sperm, and hair) are also affected, which explains the negative effects of nausea associated with chemotherapy hair loss, infertility, and others. Usually, several drugs are administered concomitantly in order to reduce the risk of resistance and to increase efficiency. Chemotherapy treatments are regularly planned cycles of therapy interspersed with segments of rest without treatment in order to give a respite to the healthy cells.

Comment #3: According to data from the medical literature (see below), however, there is presently no evidence that using chemotherapy early in combination with hormonal therapy in patients responding well to hormone therapy is the most favorable option. Additionally the patient now had an acceptable quality of life and did not wish to lose it because of the side effects of docetaxel. By taking a benefit/risk ratio approach the patient decided not to undertake this chemotherapy treatment with docetaxel and rather continue with hormone therapy at least until he became hormone resistant. At that point, he may rather opt for a second-line hormone therapy before initiating chemotherapy according to his health status.

(B) Surgical therapy. Surgery is a common treatment for prostate cancer. It is used to potentially cure the cancer by completely removing the tumor and reducing pain or symptoms (palliative treatment). The type of surgery done depends mainly on the stage of the cancer and other factors, such as the patient's age, general health condition, and life expectancy [1,2]. Side effects of surgery depend on the type of surgical procedure. Radical prostatectomy is the most common surgery to treat localized prostate cancer [1,2].

Transurethral resection of the prostate (TURP) is usually done as palliative treatment to relieve urinary obstruction. Usually, prostate cancer that has spread to the pelvic lymph nodes is considered incurable. Sometimes prostate cancer with only a microscopic spread to proximal lymph nodes can be cured with surgery. Pelvic lymph node dissection is performed during radical prostatectomy to find out if the cancer has spread to the pelvic lymph nodes.

In this procedure the major groups of lymph nodes in the pelvis are removed. In general, when the pelvic lymph nodes are involved, it is too late to treat with surgery, as in the case of our patient. Furthermore, on the abdominal scan of our patient the prostate was invading the intestinal wall, which makes it very risky to carry out surgical therapy without the risk of affecting the digestive tract. Furthermore, with metastasis in the bones and other organs, performing surgery is not a viable option. Therefore B is not the best answer.

Comment #1: Radical prostatectomy is a major surgical method and consists of a complete excision of the gland, the surrounding cells, and the seminal fluid. It would represent a better alternative than radiotherapy; the overall survival rate is especially better in patients that reached the high-risk prostate cancer stage [3]. In addition, when combined with radiotherapy and antiandrogenic therapy the overall survival rate is improved [3].

The radical prostatectomy can be performed via different approaches: retropubic, perineal, and laparoscopic [1,2]. Retropubic radical prostatectomy is done through an incision at the lower abdominal region up to the pubic bone. This makes it possible to remove the lymph nodes around the prostate if tumor cells are suspected to be housed there.

When performing a perineal radical prostatectomy the surgeon breaks the skin (perineum) between the anus and the scrotum. This approach is used when the removal of lymph nodes is not necessary; patients often prefer this method because it causes fewer erectile problems [1,2]. However, as explained earlier, because of the prostate cancer presentation of our patient, none of the surgical approaches briefly discussed here are applicable.

Comment #2: The last surgical possibility is when the surgeon performs several incisions to remove the prostate. This method is advantageous because it does not fully open the skin and there is less bleeding; also the hospitalization is shorter and the remission faster. The major side effects of a radical prostatectomy, in addition to those associated with the surgery are urinary incontinence, erectile dysfunction, changes in orgasm, loss of fertility, a change in the length of the penis, lymphedema (though rare), and an increased risk of inguinal hernia. However, as explained above, considering that our patient had widespread lymph nodes on the CT scan and many metastases on the bone scan, this approach was not applicable for him. Therefore C is not the best answer.

(C) Radiation therapy. Radiation therapy is a technique that uses waves (or rays) at a very high power and high energy in order to reduce the size of tumor (s) via the elimination of tumor cells [1,2]. This radiation interferes with the DNA of the cell directly or by creating free radicals inside the cell that will interact with the cell and cause irreparable damage [1,2]. There are two types of radiation in the treatment of cancer: an external radiation i.e., X-rays form a source of radiation from outside the body and an internal radiation (also called brachytherapy), where the radiation source is placed inside or near the tumor in the body.

In both cases, healthy cells can also be affected; even though, they are still located processing tools. The choice of the type of therapy depends on the type of cancer, its location, whether there are metastases, the health of the patient, and the treatment that the patient follows or will have to follow.

For prostate cancer, radiation therapy may be used:

- as the first therapy in low stage cancers located more inside the prostate
- as a first treatment for a tumor located in the tissue around the prostate
- in cases where the cancer has not been completely resected or is recurrent
- when the cancer is advanced in order to reduce the tumor size
- or as in our patients to relieve bone pain (palliative care) [1,2]

External radiotherapy has the objective of targeting more effectively the prostate tumor in order to reduce the exposure of surrounding healthy tissue via more specific methods such as:

- conformational radiotherapy with modulations of intensity
- conformational radiation in three dimensions
- stereotactic body radiotherapy

Generally the treatment is given 5 days a week for 7–9 weeks with a radiation of a very short duration [1,2]. However, as mentioned earlier, being that our patient had a highly invasive prostate cancer, the only option for him in terms of radiation therapy was palliative radiotherapy to reduce bone pain.

(D) Brachytherapy allows for the injection of radioactive seeds into the prostate by using a fine needle that passes through the skin between the scrotum and the anus. Radiation therapy is permanent until the grain or pellets of low radiation are withdrawn a few weeks later, or it can be intermittent via a catheter that is left in place in order to administer other short treatments over a certain period of time [1,2].

There is an excellent review of the medical literature comparing external radiation therapy, brachytherapy, and radical prostatectomy [4]. According to Wolff et al., these are all effective therapies in the localized treatment of prostate cancer. In fact, external radiotherapy seems to have a survival rate slightly more favorable than the radical prostatectomy. There are more side effects when combined with hormone therapy, but the overall survival rate is significantly better [4].

Brachytherapy in turn would ensure a better quality of life and chance of survival without disease compared to a radical prostatectomy; it would also lead to a better and more satisfying sexual function. Finally the main adverse effects associated with the two therapies in the treatment of prostate cancer are gastrointestinal, urinary, and sexual problems (erection). For the reasons mentioned above, this approach was not acceptable for the patient because he already had bone metastasis at the time of diagnosis. Therefore D is not a good answer; however, for the benefits of the reader, this approach is briefly discussed here.

(E) Hormone therapy. Hormone therapy, as the name suggests, involves the use of hormones (endogenous or exogenous) for therapeutic purposes. Hormones are molecules produced by glands of the human body [1,2]; they move through blood vessels to reach all areas of the body. When cells expressing one or more receptors for hormones, they may interact with the functioning of these cells. This is generally done via a modulation of the profile of gene expression in these cells. One of the functions of hormones, which is crucial in the treatment of tumors, is the control of the development and the growth of some types of cancerous cells responding to hormones [1,2].

The hormone therapy suggested to our patient consists in eliminating the male hormones from his body and included: a hormone LH-RH agonist (luteinizing hormone-releasing hormone); which is a hormone produced by the hypothalamus, located at the brain base [1,2]. This hormone controls the secretion of sex hormones. These types of drugs bind to LH-RH receptors of the hypothalamus and cause a transient hypersecretion of hypothalamic hormones of LH and FSH. This hypersecretion may increase the circulating levels of testosterone, causing a transient upsurge of the symptoms, especially bone pain.

After this transitional episode, called the rebound effect, there is a drying up of the FSH and LH secretion, which results in a decreased level of testosterone. The hormone therapy also included nonsteroidal antiandrogen drugs, which have the goal of blocking the androgenic receptors, therefore making impossible the action of male hormones on the androgenic receptors [1,2]. Given the side effects of Casodex and risks of resistance to hormone therapy, the patient opts for intermittent androgen suppression (IAS) rather than continuous androgen suppression [5].

Comment #1: This hormonal therapy consists of a nonsteroidal antiandrogen such as Casodex 50 mg administered orally once a day (the first dose was given on August 8, 2014) and 10.8 mg Zoladex (a LH-RH agonist, an analogue of gonadotropin-releasing hormone) by subcutaneous injections every 3 months (the first dose was received on August 11, 2014); the latter dosage should be given for the rest of the patient's life. Considering the benefit/risk ratio of this treatment, the patient decided to go for the first line hormone therapy.

Comment #2: The target of IAS is based on the possibility of delaying a tumor hormone-resistant clone selection and improving treatment tolerance. In this technique the treatment is stopped after a certain time or after the collapse of the PSA. When the PSA goes back, treatment is resumed; it can be halted again if the PSA level normalizes [5]. In addition, according to the Canadian Product Monograph for Casodex, some patients with metastatic prostate cancer, the Casodex may stimulate the growth of prostate cancer cells rather than inhibit it. A decline in PSA, a clinical improvement, or both may be reported after the withdrawal of the antiandrogens. For more information about this product, please see: https://www.astrazeneca.ca/content/dam/azca/downloads/productinformation/CASODEX%20-%20Product%20Monograph%202016-11-17.pdf.

Comment #3: In the case of our patient, it was recommended to discontinue treatment with Casodex because his PSA was very low (<0.1), as well as to follow the patient for 6–8 weeks in order to detect any response to the withdrawal of the antiandrogen before making the decision to adopt another form of treatment. The patient continued on IAS for a period of 2½ years because his PSA remained very low (<0.2). Furthermore the choice of the patient to opt for IAS is supported by medical literature [5]. In addition, these types of drugs are associated with significant side effects, including tiredness, a decreased sex drive, hot flashes, and changes in body appearance (e.g., a favor of an increase in fat mass and a decrease in muscle mass). After 2½ years of IAS the patient's PSA is still low (<0.2), but it has increased significantly to 0.9 over a period of 2 months; thus Casodex was reintroduced on June 5, 2017.

Comment #4: Zoladex is also associated with the risk of osteoporosis, for which the patient should take 1000 mg of calcium and 800 international units of vitamin D for the rest of his life. To help with the treatment of osteoporosis, the oncologist also suggested Xgeva at 120 mg every 6 months. Xgeva (denosumab) is a fully human IgG2 monoclonal antibody with a high affinity and specificity for human RANK Ligand (RANKL). The binding of Xgeva to RANKL inhibits RANKL from activating its only receptor, RANK, on the surface of osteoclasts and their precursors. Increased osteoclast activity stimulated by RANKL is a key mediator of bone disease in metastatic tumors. The prevention of RANKL-RANK interaction inhibits osteoclast formation, function, and survival, thereby decreasing bone resorption and interrupting cancer-induced bone destruction.

Xgeva is also associated with significant side effects, including avascular necrosis of the jaw when dental procedures are performed, hypocalcaemia, and spontaneous risk fracture. In the absence of a demonstrated effectiveness of this medication in patients sensitive to hormone therapy and the presence of significant side effects, the patient delayed taking this drug (see Case Report #6). In addition, there is a low risk of fracture related to osteoporosis.

Fortunately, this was confirmed during a tomodensitometry bone scan performed on September 5, 2014, which concluded that no pharmaceutical treatment for osteoporosis was required at this time. Our patient must also take Asaphen 80 mg/day (children's aspirin) to prevent the possibility of myocardial infarction, another side effect of Zoladex.

(F) Surgical castration. This is also known as orchiectomy or orchidectomy and entails the removal of a man's testicles [6]. Surgical castration is a type of prostate cancer hormone therapy that is nonreversible. Historically, surgical castration was the only type of hormone therapy available for the treatment of prostate cancer; therefore hormone therapy was used only after every other treatment option had been exhausted. Orchiectomy is still used today due to both its low cost and limited number of side effects. Orchiectomy is less expensive than chemical castration through hormone therapy. Many men also find that the side effects of orchiectomy are also fewer and less severe than those of chemical castration. The oncologist suggested the following treatment options: either a physical castration or a chemical castration with hormone therapy. The patient was not enthusiastic about the physical castration option and so he selected a chemical castration with hormone therapy. Therefore D is the best answer.

Discussion: Fortunately the patient responds very well to hormone therapy. After the first 10 months of treatment with Zoladex and Casodex, his PSA went down from 173 to 0.6 (normal is between 0 and 4), then to 0.2 and 0.1 few months later; it then returned and stabilized to 0.2 for 2½ years. Once the patient was stabilized at a PSA level <0.2 for at least 6 months the decision was made to remove Casodex and follow an IAS approach (see above).

The CT scan performed on March 12, 2015, showed that the overwhelming majority of lung metastases disappeared and that abdominal, pelvic, and other lymph nodes became normal size, with the disappearance of the lymph nodes obstructing the right urinary tract. The right kidney is back to normal volume, while the right kidney hydronephrosis disappeared; with it the urinary functions came back to normal, which was very reassuring for the patient. As mentioned before, considering that this patient was still having a PSA < 0.2 after 2½ years of hormone treatment, it is likely that his life expectancy will be much longer than the statistical data mentioned above (see Case Report #1).

However, it is important to note that despite the patient's medical knowledge, contacts, and insistent efforts the path between the appearance of the first symptoms (November 2013) and the start of the patient's treatment (August 2014) was long. Therefore it is highly recommended that patients remain active and vigilant in during treatment, as a passive approach is risky and will most probably result in delays and poor results. It is important to shake up the slow pace of the medical system and get faster treatment. The patient's responsibility is to ensure that they are compliant with their treatment and follow-ups.

REFERENCES

[1] Canadian Cancer Society. What is prostate cancer? Available at: http://www.cancer.ca/en/cancer-information/cancer-type/prostate/prostate-cancer/?region=on; 2017.

[2] American Cancer Society. What's new in prostate cancer research? Available at: https://www.cancer.org/cancer/prostate-cancer/about/new-research.html; 2017.

[3] Lei JH, Liu LR, Qiang Wei Q, et al. Systematic review and meta-analysis of the survival outcomes of first-line treatment options in high-risk prostate cancer. Sci Rep 2015;7713:1–9.

[4] Wolff RF, Ryder S, Bossi A, et al. A systematic review of randomised controlled trials of radiotherapy for localised prostate cancer. Eur J Cancer 2015;51:2345–67.

[5] Harvard Medical School. Intermittent hormone therapy for prostate cancer. Harvard Health Publications; 2017. Available at: http://www.harvardprostateknowledge.org/intermittent-hormone-therapy-for-prostate-cancer.

[6] Hackethal V. Surgical castration instead of drugs in prostate cancer. Medscape Family Medicine 2015. Available at: http://www.medscape.com/viewarticle/856574.

CHAPTER 4

Case Report #4—External and Internal Risk Factors

INTRODUCTION

With the exception of his prostate cancer this patient is otherwise in good health, without a relevant medical history or positive family history of cancer. He is a physically active man who eats very well. He is a nonsmoker, does not consume alcohol or drugs, but he does have a history of exposure to environmental contaminants during his early adolescence. According to the WHO, about 30% of cancer cases are preventable [1]. To be successful in our prevention, we must target the main risk factors associated with cancer. It is possible to categorize these risk factors into two main groups: external factors that are possible to treat but sometimes difficult to control and internal factors that cannot be influenced or changed. Considering that the internal risk factors cannot be changed, we will discuss the external risk factors we can work on with the goal of helping our patient and any other patients suffering from cancer to better understand the origin or cause for their cancer.

Question #1: With the information presented above, what external risk factors should we consider to better understand the origins of this patient's prostate cancer?

(A) Tobacco smoking
(B) Alcohol consumption
(C) Obesity
(D) Sedentary lifestyles
(E) Dietary habits
(F) Infections
(G) Environmental pollution and occupational exposure to carcinogens
(H) Radiation exposures

Answer: G.

Prostate Cancer
https://doi.org/10.1016/B978-0-12-815966-8.00004-7
© 2018 Elsevier Inc.
All rights reserved.

(A) Tobacco smoking. The WHO estimates that tobacco would be responsible for about 22% of cancer-related deaths annually. The prevalence of smoking [1] in Canada is 19% for individuals aged 12 and older [2]. This risk factor undoubtedly has the strongest correlation to cancer, particularly lung cancer, though it is also a large risk factor in about 17 other types of cancer, including colorectal cancer, kidney, liver, and ovary [3]. It is therefore not surprising that several countries are investing time and money in the fight against smoking. According to the Canadian Coalition for Action on Tobacco   (http://www.smoke-free.ca/eng_home/ccat-july2002-francais.htm), in 2014, Health Canada spent $38 million in various strategies to reduce the rates of disease and death associated with tobacco smoking [3]. These efforts seem to bear fruit because a decrease in the incidence rates of smoking-related cancers is expected for 2028–32 [4]. As our patient is a nonsmoker and was not much exposed to passive smoking, his prostate cancer is unlikely associated with smoking.

(B) Alcohol consumption. Tobacco and alcohol are the main external risk factors and when consumed together, their effect potentiates the risk of certain cancers [2–5]. Prevention recommendations made by the government consist on two drinks for women and three drinks for men per day without consuming on a daily basis in order to reduce the long-term negative effects associated with alcohol consumption aside from cancer [5]. As our patient was not drinking much alcohol (one or two glasses of red wine per week), B is unlikely to be a major risk factor for him.

Comment #1: Many people, however, are not aware that alcohol is a serious carcinogenic. Indeed, alcohol consumption increases the risk of developing cancers such as breast, colorectal, esophageal, larynx, oral cavity, and liver cancers [5]. This means that the risk of developing cancer increases in parallel with the increase of alcohol consumption [5].

Comment #2: However, it is confusing when considering alcohol consumption guidelines in the prevention of cancer. This guideline suggests that consuming not more than two drinks for men and one drink for women per day can prevent 90% and 50% of cancers linked to alcohol over-consumption in men and women, respectively [5]. Even a slight increase in the risk of breast cancer has been associated with the consumption of three to six drinks per week [5]. As for smoking, Health Canada is involved in the prevention of harm associated with alcohol consumption through agencies such as the Canadian Center on Substance Use and Addictions. For more information regarding this agency, please consult the following link: http://www.cclt.ca/eng/pages/default.aspx. In fact, according to an analysis performed in 2012, $20.3 billion was spent in Canada for the period 2010–11 for the prevention of alcoholic beverages [6].

Comment #3: Additionally the Canadian Alcohol and Drug Use Monitoring Survey (CADUMS), whose last highlights are available via the following website: https://www.canada.ca/en/health-canada/services/health-concerns/drug-prevention-treatment/canadian-alcohol-drug-use-monitoring-survey.html and with whom the Canadian Centre for Substance Abuse (CCSA) worked, reported that in 2012, 78.4% of Canadians stated that they are drinking alcohol. Once again, we notice the important role played by different regulatory agencies, such as Health Canada and its partners to prevent extreme alcohol consumption. This can certainly help to reduce the risk of cancer associated with alcohol overconsumption. For more information about the CCSA, please consult the following link at: https://www.ccsa.cable.ca/.

(C) Obesity. The next three factors, obesity, sedentarity, and diet, are interrelated and are considered to be a combined integral part of the risk factors that can be modified to reduce the chances of developing cancer [7,8]. In order to categorize the weight of a person on a common basis, the body mass index (BMI) is used. This measure takes into account the weight and height of an individual to give an index, which according to the result can be tabulated in low (BMI $\leq 20 \, \text{kg/m}^2$), normal ($20 < \text{BMI} < 25 \, \text{kg/m}^2$), weight in excess (BMI $\geq 25 \, \text{kg/m}^2$), or obese (BMI $\geq 30 \, \text{kg/m}^2$) [7]. According to Statistics Canada, in 2014, 61.8% of men and 46.2% of women and 23.1% of adolescents between 12 and 17 years of age have been characterized as having excess weight or being obese [8]. This is a large number of people at risk of developing several pathologies. For more on the prevention and treatment of obesity, see a recent article written by Plourde G and Prud'homme D on the management of obesity [8].

Overweightness or obesity is a risk factor associated with breast, cervical, ovarian, colorectal, esophagus, kidneys, liver, pancreas, and prostate cancers [2]. On the other hand, a low body weight is correlated to, among other things, esophageal cancer. However, few studies are directed towards low weight and cancer risk; therefore it is essential to maintain an ideal body weight to lower the possibilities of having cancer. Considering that our patient was of normal weight, it is therefore unlikely that obesity was a risk factor for him.

(D) Sedentary lifestyle. The benefits of regular physical activity continue to be promoted, as well as the adverse effects associated with a sedentary lifestyle [9]. These benefits are even more important when people are more active. It is estimated that 51% of adults in the United States and 31% of adults in the entire world aren't doing the minimum amount of exercise recommended for good health. The guidelines on physical activity needs vary according to age groups. According to Statistics Canada, in 2014, 53.7% of Canadians over 12 years old reported that they were indulging in physical activities during their leisure time, and 54.0% of people over 18 years old are considered to be overweight or obese [8,10]. It is currently recommended for adults between 18 and 64 years to do at least 150 minutes of exercise of moderate to high intensity per week. This can be done by periods of at least 10 minutes with a few sessions of muscle reinforcement [8,10]. The exercise can include walking, running, swimming, biking, or other activities that you enjoy with the goal of achieving the recommendations mentioned above [8,10]. As our patient was very active, it is unlikely that this risk factor was associated with the development of his prostate cancer (see Case Report #8 for further discussion on the role of physical activity in improving the quality of life of patients with prostate cancer).

Comment #1: In a recent study of 1.44 million participants aged 19–98 in the United States and Europe followed for an average of 11 years. During that period, 187,000 new cases of cancer were diagnosed. On the other hand, researchers have detected a reduction in the risk for the following cancers with regular physical activity: esophageal (−42%), liver (−27%), lung (−26%), kidney (−23%), stomach (−22%), endometrium (−21%), blood (−20%), colon (−16%), and breast (−10%) [11]. In most cases the link between regular physical activity and the decline in cancer risk is continuously observed, regardless of the person's weight or whether he is a smoker. This study confirmed the link already highlighted between the beneficial effect of exercise and the large decline in the risk of cancers. For all cancers, the risk reduction resulting from regular physical activity was around 7% [11].

Comment #2: On the other hand, physical activities have been linked to a 5% increase in the risk of prostate cancer and a 27% risk of melanoma, an aggressive cancer of the skin, especially prevalent in very sunny areas of the United States [11]. However, regular physical activity helps maintain a healthy weight, which also minimizes most risks of cancers related to obesity and to smoking [1,2,12]. Obviously the benefits of doing regular physical activity largely outweighed the risk for cancer.

(E) Dietary habits. Diet is, of course, complementary to physical activity and is therefore a contributor to the maintenance of a healthy weight, thus reducing the cancer risk that is associated with overweightness or obesity.

Some food habits are correlated with the prevention of the onset of cancer, or at least in avoiding weight gain. These habits include an appropriate consumption of fruits, vegetables, and fibers, as well as a low consumption of red meat or processed food, a reduction of fast food, salt, fat, and sugar [12,13]. Again, based on this case report, it is unlikely that diet was a risk factor in our patient (see Case Report #8 for further discussion on the role of diet in the prevention and treatment of prostate cancer).

Comment #1: To date, the Canadian Food Guide recommends for an adult 7–10 servings of fruits and vegetables as well as 2–3 servings of meat and substitutes per day. There is no reason to believe that these recommendations are different for patients with cancer. In 2014, only 39.5% of Canadians reported eating >5 servings of fruits and vegetables per day [14]. Despite the efforts made by the Government of Canada and other national and international organizations in the promotion of healthy eating and healthy lifestyles, this proportion remains low. According to data collected the following cancers are suspected of being linked to poor diet: stomach, esophageal, colorectal, and oral cavity [15].

It is the individual responsibility to make the decision to adopt a healthy diet and regular physical activity schedule in order to maintain a healthy weight and avoid a sedentary lifestyle, thus reducing the risks associated with the development of different types of cancers.

(F) Infections. According to the WHO, infectious agents would be responsible for deaths by cancer for about 6% in industrialized countries and 22% in developing countries [1]. The relationship between infections and cancer is rather paradoxical: People who face infections are often more predisposed to develop cancer in the same way that people with cancer are more likely to get infections. A weakened immune system can increase the risk of infection and/or cancer [16]. The main agents that may be associated with the onset of cancer are usually viral, bacterial, or parasitic in nature [17]. Obviously, our patient was exposed to various infectious agents during his life, but he did not have any significant infection that could be related to the development of his prostate cancer. Furthermore, our patient was assessed for a viral syndrome, and the screening came back negative (see Case Report #2). Therefore F is not the best answer.

Comment #1: Viral hepatitis B and C (HBV and HCV), the Epstein-Barr virus (EBV) the human type 1 T lymphotrope virus (VLTH-1), the human herpes 8 virus (VHH-8), and the human immunodeficiency virus (HIV) are viruses known to be related to certain types of cancers [17]. Again, our patient was tested for these types of infections, and none of them came back positive. This confirms that his advanced prostate cancer is not at all related to the above types of infections.

Comment #2: EBV, VLTH-1, VHH-8, and HIV are among other viruses that were tested, as they are the major viruses associated with lymphoma [17]. The latter was considered to be an initial differential diagnosis in our patient. However, after investigation, these viruses were considered negative, and the diagnosis of lymphoma was also excluded (see Case Report #2).

(G) Environmental pollution and occupational exposure to carcinogens. Several environmental agents polluting the air, soil, or water have already been studied for their possible links to different types of cancers. Worldwide, about 19% of all cancers could be explained by exposure to environmental factors, including occupational exposures [18]. Second-hand smoke, burning wood or coal, and carbon monoxide are good examples of indoor and outdoor air pollution.

Occupational exposures to different types of carcinogenic are another form of environmental pollution. Indeed, those who exposed themselves in a more deliberate and constant way to some carcinogens because of their professional practice had a higher risk of pathological diseases such as cancer. According to the WHO the most common profession-related cancers are lung cancer, mesothelioma, and bladder cancer [1]. Professional carcinogenic agents include, for example, asbestos and formaldehyde, both of which are used in the manufacturing of some building or textile materials, as well as in some household products such as paint (formaldehyde mainly) [1,2]. As stated above, during his early adolescence, our patient was exposed to pesticide, herbicides, and fungicides. Even though, we have no way to confirm a true causal association between these exposures in his early adolescence and the development of his prostate cancer, a link is documented in the medical literature. Therefore G is a correct answer.

Comment #1: The prevention of cancers related to these various environmental pollutants begins with a limitation in the exposure to situations where these pollutants are predominant. Professionals, for example, must submit to the guidelines established by the employer and those from the Health and Safety at Work Commission (Canadian Center for Occupational Health and Safety [CCOHS]). For more information regarding this latter organization, please consult the following link: https://www.ccohs.ca/. We must also try to eliminate the pollution source if possible (e.g., second-hand smoke) or ventilate the site of exposure when using fuel sources. Subsequently, as with all the cancer risks, early detection is an effective way to reduce the chances of developing the disease. Given the growing number of environmental agents listed that may have an effect on public health, it is important to be able to identify them. Many people do not know all the daily elements to which they are exposed that can make them more likely to be at risk of certain types of cancers. Thus, it is crucial to consult different sources of information, such as the websites of the International Center for Research on Cancer (ICRC, WHO agency); the United States Environmental Protection Agency: https://www.epa.gov/newsroom/browse-news-releases; the Occupational Cancer Research Center: http://www.occupationalcancer.ca/; and Health Canada in order to be well informed of the consequences, treatments, and overall prevention available. These sites present an exhaustive list of environmental factors to which people may be exposed. To view these sites, consult the Resource for Partners section.

(H) Radiation exposures. Skin cancer is without a doubt the most common cancer associated with this risk factor. The emphasis placed on the prevention of this type of cancer has increased in recent years. The awareness of the benefits of reduced exposure to the sun or to tanning beds, the preventive use of sunscreens, and the importance of wearing adequately protective clothes while exposed to the sun are examples of simple and effective ways to prevent and reduce this type of cancer. However, what most people don't know is that there are two kinds of radiation: ionizing and nonionizing [18]. Ionizing radiation is associated with a higher risk of skin cancer, but it does not seem to be related to prostate cancer; therefore H is not the best answer.

Comment #1: The first category of radiation includes radon and medical radiation, ambient radiation (cosmic or radioactive material) as well as radiation from human activities (weapons or nuclear energy). The second category includes electromagnetic fields (EMF) and radio frequencies (RF) related to various technologies, such as cell phones and appliances [2]. For the latter, several studies should be conducted regarding their causal relation to cancer. However, the Canadian Institute for Research in Cancer (CIRC) seems to believe in a link between EMF/RF and cancer [19]. The fact that skin cancer is strongly linked to ultraviolet rays (UV) is also part of this second group. According to the WHO, in 2002 the UV rays have been associated to 60,000 cases of cancer, including 48,000 cases of melanoma. In Canada, a 72% increase in new cases of melanoma is expected between now and 2028–32 [20].

Comment #2: UV-rays, X-rays, and gamma rays are also linked to breast, prostate, central nervous system (CNS), and stomach cancers [20]. Radon is another important carcinogenic source that can be found in the air, the ground, and in some building materials. According to the WHO, it would be the second risk factor for lung cancer. Of course, there are other radiation agents that must be clarified, but like the polluting agents, the best prevention is to limit or avoid exposure to such radiation, to test for early detection and to raise awareness of the effects generated by this factor. It is true that our patient was exposed to UV rays, X-rays, and gamma rays during his life, as he was working a good deal outside the house during the weekend and vacation. Similarly, during his training in medicine and for his PhD, he was exposed to various radioactive products in order to perform fundamental research in drug development. At this time, it is impossible to confirm a link between his prostate cancer and these risk factors. Therefore, G is still the best answer.

Question #2: True or False? The internal factors play no role for our patient in the development of his prostate cancer.

Answer: False.

In the previous section, we spoke mostly about external factors, for which it is possible to a certain extent to intervene. In this short answer, we will talk briefly about the internal factors for which we have no influence. Although cancers can develop at any age, they are much more common after the age of 60. Our patient was diagnosed with an advanced prostate cancer at 52, which is not in the main age group for this kind of cancer. However, due to the accumulation of external attacks suffered by the cells and probably to the lower efficiency of the DNA repair mechanisms with age, we cannot totally exclude the age factor [20].

There is a range of measures to refer to in order to screen the different types of cancer according to age. For more information on the screening for prostate cancers, see the links from the Canadian Task Force on Preventive Health Care [21] in the Resources for the Readers section.

Heredity may also play a role. Some people are more likely to develop cancer than others because they have mutations in one or more of their genes since birth. These are inherited from their relatives and are present in all of their cells [22]. It seems that the cells of those people had already taken several steps that could lead to the process of cancerization; only a few mutations acquired later in adult life are then required to activate the process.

When this type of mutation is involved in cancer, one speaks of a hereditary form or genetic predisposition to cancer. Only 5%–10% of cancers are related to the transmission of a known hereditary mutation; in most cases, they are associated with a very evocative family history of cancer, which is not the case for our patient. The best-known hereditary mutations predisposed to cancer are the BRCA1 and BRCA2 genes that produce a significant increase in the risk of breast and ovarian cancers. People with this mutation have a 50% chance to transmit it to each of their children. There are three broad categories of genes associated with cancer pathologies: the oncogenes, tumor suppressor genes, and DNA repair genes [1,2]:

- The oncogenes are genes whose expression promotes the development of cancers. These are the genes that control oncoprotein synthesis (i.e., proteins stimulating cell division) and trigger a disorderly cell proliferation.
- The second category includes the suppressor genes in tumors that are negative cell proliferation regulators. Unlike the oncogenes that become hyperactive in cancer cells, tumor suppressor genes lose their function in human cancers.
- Finally, there are also repair genes that are able to detect and repair damage to DNA, which changed the oncogenes or tumor suppressor genes. It also happens that these repair systems are inactivated by the cancer cells.

Cancers are genetic diseases; that is, they have caused a quantitative and/or qualitative change in the genes described above. As cancer represents somatic genetic alterations that are present in the diseased tissue, most cancers are therefore not hereditary [1,2]. For our patient, we are tempted to eliminate a hereditary form of cancer, considering he has no family history of prostate cancer; however, we cannot eliminate a nonhereditary form. That is why we cannot totally eliminate a role of internal factors in the development of his cancer.

Question #3: True or False? In the case of prostate cancer, early diagnosis and treatment has no influence on its development.

Answer: False.

Early detection is very important for the following reasons:
- For an early diagnosis which consists of being aware of the early signs and symptoms of cancer in order to diagnose and treat the cancer early.
- Early diagnosis is particularly relevant when there is no efficient method of screening. In the absence of early detection, patients are diagnosed very late often when a curative treatment is no longer possible.

- By aiming to identify and test abnormalities suggestive of a particular cancer at a precancerous stage in a patient and then refer quickly to the qualified medical resource for diagnosis and treatment, we increase our chances of success.

For our patient the diagnosis and treatment of his prostate cancer were performed too late.

Screening programs are particularly effective for common cancers, for which available tests are accessible to the majority of the exposed population. Here are some examples of screening:

- Visual inspection after the application of acetic acid (IVA) for cancer of the cervix
- Test for HPV for cancer of the cervix
- Pap test for the detection of cervical cancer
- Mammography for screening of breast cancer
- PSA level, especially when the doubling time is short, and DRE have important roles in the screening of prostate cancer

Of course the accuracy of the diagnosis is essential for appropriate and effective treatment because each type of cancer requires a specific protocol that includes one or several treatment modalities like surgery, radiation therapy, chemotherapy, or hormone therapy, as discussed previously.

The most important goal of treating prostate cancer is to cure the disease or prolong life and increase the quality of it. Unfortunately, there is not yet a cure for metastatic prostate cancer; therefore the main goal for our patient is to prolong his quality of life. Improving the quality of life for patients with this disease is also a major objective. Supportive care in the form of an adequate diet, regular physical activity, and stress management can certainly help. This will be further discussed in the Case Report #8. Obviously, healing can be improved, if cancers are detected and treated early in accordance with the best Clinical Practice Guidelines. This was even more evident for our patient, who presented with a very aggressive and intrusive form of prostate cancer from the onset; this suggests that he would have benefited from a more rapid detection and treatment instead of losing many months on treating an unlikely condition, such as chronic prostatitis.

Discussion: Despite the progress that has led us to better understand the mechanisms of cancer development the causes of prostate cancer are currently not fully known. As mentioned earlier, we have identified risk factors that appear related to prostate cancer, with age being the biggest risk factor. Prostate cancer is not common before the age of 55, and it is most often discovered after the age of 70.

In some cases, prostate cancer seems to have a very strong family association [1,2]. According to family history, prostate cancer can occur in three forms:

- The sporadic noninherited form (this is the most common form)
- The familial form, which involves at least two cases of prostate cancer among related parties of the first degree (e.g., father or brother) or secondary degree (e.g., grandfather or uncle). This familial form represents 20% of prostate cancers diagnoses
- The hereditary form, which is defined by the existence of at least three cases of prostate cancer among relatives in either the first or second degree or two family members diagnosed before the age of 55. The hereditary form represents 5% of prostate cancers.

Studies are underway to identify genetic mutations promoting the risk of prostate cancer. To date, several genes showing predisposition to prostate cancer have been studied, but no evidence is conclusive about their susceptibility to interact. The role of genes in the development and treatment of cancer (i.e., personalized medicine) will be discussed more deeply in the Case Report #8.

Ethnic and geographic origins must also be considered. Indeed, many studies have shown that the number of prostate cancer diagnoses is higher in Northern Europe and North America while it is lower in Southeast Asian countries. It was also established that Afro-Caribbean men have a higher risk of developing prostate cancer [1,2].

It is also possible that all the patient's life conditions may contribute to the development of prostate cancer [15], but the analysis of these factors is extremely complex. For example a large consumption of milk and dairy products can contribute to a high input in calcium, which is likely associated with an increased risk of prostate cancer, but this needs to be confirmed [22,23].

Researchers also focused on various medical factors that could constitute risk factors or risk markers [24]. For example the prostatic intraepithelial neoplasia (PIN) is one of these factors. It is a type of precancerous cell proliferation that can sometimes turn into prostate cancer. The PIN is divided into two categories: low levels and high levels. The PIN of high levels may represent an intermediate stage between a benign tumor and a malignant tumor. Men diagnosed with a high PIN after a prostate biopsy as in our patient could experience a higher risk of prostate cancer and therefore must be monitored closely via prostate cancer screening [1,2]. But for our patient the PIN has no value because at the time of biopsy, he was already diagnosed with an invasive prostate cancer.

In summary, it is difficult for our patient to accept this aggressive and intrusive prostatic adenocarcinoma for several reasons. First, this patient is not in the usual age group and has no family history of prostate cancer or other cancers. He does not have clearly identified external factors, either. The prostate examination conducted 5 years before showed no benign enlarged prostate or other abnormality. We are therefore tempted to say that the risk factors associated with the development of his prostate cancer are most likely multifactorial; that is, a combination of genetic and environmental factors.

It would be unwise for the author to minimize the importance of what has been discussed above because it appears that in some cases cancer can develop without apparent risk factors. In this case the early detection and treatment are still the only effective way of preventing further development of prostate cancer. Of course, as our patient is the victim of prostate cancer at a relatively early age the risk of prostate cancer for his brothers is therefore significantly increased; hence prevention and early detection of prostate cancer become very important for them.

To facilitate the detection of cancer including prostate cancer, it is recommended for the health care providers to follow the Clinical Practice Guidelines, including those from the Canadian Task Force on Preventive Health Care. Some of these guidelines are presented in the resources for readers section. These Clinical Practice Guidelines are also important for all other patients at risk, regardless their ethnic origin and whether they are associated with different risks of cancer, as mentioned above.

REFERENCES

[1] World Health Organisation. Prevention of cancer. Available at: http://www.who.int/cancer/prevention/en/; 2017.

[2] Canadian Cancer Society. What is prostate cancer? Available at: http://www.cancer.ca/en/cancer-information/cancer-type/prostate/prostate-cancer/?region=on; 2017.

[3] Canadian Coalition for Action on Tobacco. A brief submitted to the House of Commons Standing Committee on Public Safety and National Security. Available at: http://www.cqct.qc.ca/Documents_docs/DOCU_2008/MEMO_08_05_05_CCAT_ContrabandHearings.pdf; 2005.

[4] Canadian Cancer Society's Advisory Committee on Cancer Statistics. 2017. Available at: http://www.cancer.ca/en/cancer-information/cancer-101/canadian-cancer-statistics-publication/?region=bc.

[5] Pelucchi C, Tramacere I, Boffetta P, et al. Alcohol consumption and cancer risk. Nutr Cancer 2011;63:983–90.

[6] Thomas G. Analysis of beverage alcohol sales in Canada alcohol price policy series, report 2 of 3; November 2012. Available at: http://www.ccsa.ca/Resource%20Library/CCSA-Analysis-Alcohol-Sales-Policies-Canada-2012-en.pdf.

[7] Health Canada. Canadian guidelines for body weight classification in adults. Available at: https://www.canada.ca/en/health-canada/services/food-nutrition/healthy-eating/healthy-weights/canadian-guidelines-body-weight-classification-adults/questions-answers-public.html; 2017.

[8] Plourde G, Prud'homme D. Managing obesity in adults in primary care. CMAJ 2012;184:1039–44.

[9] Plourde G, Marineau JM. Reducing sedentary behaviors in obese patients. If they won't exercise more—try to make them sit less. Parkhurst Exchange; November/December 2011. p. 63–71.

[10] Canadian Society of Exercise Physiology (CSEP). Canadian physical activity guidelines. Canadian sedentary behaviour guidelines. Your plan to get active every day. Available at: http://www.csep.ca/cmfiles/guidelines/csep_guidelines_handbook.pdf; 2017.

[11] Moore SC, Lee IM, Weiderpass E, et al. Association of leisure-time physical activity with risk of 26 types of cancer in 1.44 million adults. JAMA Intern Med 2016;176:816–25.

[12] Stewart BW, Wild C, International Agency for Research on Cancer, World Health Organization. World cancer report 2014. Lyon (France)/Geneva (Switzerland): International Agency for Research on Cancer, WHO Press; 2014.

[13] The World Cancer Research Fund (WCRF)/American Institute for Cancer Research (AICR). Policy and action for cancer prevention food, nutrition, and physical activity: a global perspective' report from the World Cancer Research Fund. Available at: http://www.aim-digest.com/digest/members%20over%20yr/Policy%20aND%20ACTION%20.pdf; 2009.

[14] Health Canada. Canada's food guide. Available at: https://www.canada.ca/content/dam/hc-sc/migration/hc-sc/fn-an/alt_formats/hpfb-dgpsa/pdf/food-guide-aliment/print_eatwell_bienmang-eng.pdf; 2017.

[15] Mondo DM, Roelh KA, Loeb S, et al. Which is the most important risk factor for prostate cancer: race, family history, or baseline PSA level? J Urol 2008;179:148.

[16] American Cancer Society. Infection that can lead to cancer. Available at: https://www.cancer.org/cancer/cancer-causes/infectious-agents/infections-that-can-lead-to-cancer.html; 2014.

[17] Canadian Cancer Society. Virus and bacteria. Available at: http://www.cancer.ca/en/prevention-and-screening/be-aware/viruses-and-bacteria/?region=qc; 2017.

[18] Canadian Cancer Society. Harmful substances and environmental risks—radiation. Available at: http://www.cancer.ca/en/prevention-and-screening/be-aware/harmful-substances-and-environmental-risks/?region=on; 2017.

[19] Institute of Cancer Research. 2017. Available at: http://www.cihr-irsc.gc.ca/e/12506.html.

[20] Public Health Agency of Canada. Health promotion and prevention of chronic disease in Canada. Res Policy Pract 2015;35(Suppl. 1):200.

[21] Bell N, Connor Gorber S, Shane A, et al. Canadian Task Force on Preventive Health Care. Recommendations on screening for prostate cancer with the prostate-specific antigen test (guidelines). CMAJ 2014;186:1225–35.

[22] Canadian Cancer Society. Healthy body weight. Available at: http://www.cancer.ca/en/prevention-and-screening/live-well/nutrition-and-fitness/healthy-body-weight/?region=qc; 2017.

[23] World Health Organisation. Global strategy on diet, physical activity and health. Available at: http://www.who.int/dietphysicalactivity/Indicators%20English.pdf; 2015.

[24] Gaudreau PO, Stagg J, Soulières D, Saad F. The present and future of biomarkers in prostate cancer: proteomics, genomics, and immunology advancements. Supplementary issue: biomarkers and their essential role in the development of personalised therapies (A). Biomark Cancer 2016;8(Suppl. 2):15–33.

CHAPTER 5

Case Report #5—Paraneoplastic Syndromes Associated With Prostate Cancer

INTRODUCTION

At the end of March 2015 the patient developed strange and annoying symptoms of persistent vertigo that lasted all day and were accompanied by nausea and severe headaches that did not respond adequately to NSAIDs. On April 12, 2015 the patient visited the Emergency Department for his symptoms, and a CT scan of the head, thorax, and abdomen was performed. The CT scan of the abdomen showed the appearance of new metastases at the level of the pelvis, the pubic area, and on a few lumbar vertebral bodies that were absent in the previous bone scan survey. The CT scan of the head shows the appearance of two new metastatic lesions of 1.8 cm, located on the right frontoparietal projection and the temporal left skull.

Otherwise, there was no evidence of visible intracerebral metastatic focus.

To continue the investigations of his persistent vertigo, on August 28, 2015 the patient passed an electronystagmogram (ENG) to confirm if his vertigo was of central or peripheral origin. In short the ENG images the nystagmus (jerky movements of the eyes due to some neurological disorders) in the form of a chart that records the electrical potential differences caused by these movements. It is used to identify the origin of vertigo when a clinical examination is not evocative enough to explore the balance disorders. The same day the patient had a lumbar puncture evaluating for the presence of cancer cells in his spinal fluid (i.e., liquid surrounding the central nervous system, CNS). During this puncture, it should be noted that the liquid was perfectly clear and comparable to normal CNS liquid, which goes against the presence of cancer cells in the CNS. The patient was advised to take Gravol or other comparable antivertigo medication such as Serc, scopolamine patches, and others. However, the use of these drugs has been unsuccessful, causing drowsiness in the patient and reductions in his

© 2018 Elsevier Inc.
All rights reserved.

concentration. For these reasons, it was decided that the patient would stop taking these families of antivertigo drugs. The patient also had eye exams by an ophthalmologist, who did not find any elements that could explain his vertigo from an eye perspective.

Question #1: Considering the information presented above, what should be the probable cause of these persistent vertigo and headaches?

(A) Bone metastasis in the brain
(B) Benign positional vertigo
(C) Disembarkment syndrome
(D) Para neoplastic syndrome affecting the cerebellum
(E) Vestibular migraine
(F) Systemic inflammatory response syndrome
(G) Autoimmune reaction

Answer: D.

(A) Bone metastasis in the brain. Bone metastasis is a cancer that started in another part of the body (primary) and spread to the bone. It is sometimes called secondary bone cancer or metastatic bone disease [1,2]. Bone metastasis is not the same as cancer that starts in the bone (i.e., primary bone cancer). Bone metastases are much more common than primary bone cancer; some kinds of cancer are more likely to spread to the bones than others. The most common types of cancer that spread to the bones are breast, prostate, lung, kidney, and thyroid [1,2]; however, cancer can spread to any bone in the body.

The most common sites of bone metastases in prostate cancer are the vertebrae (bones of the spine), ribs, pelvis (hip bone), sternum (breastbone), and skull, as experienced by our patient. Sometimes only one bone area is affected, while metastases can develop in several bones at the same time. Bone is constantly being formed and broken down. This is a normal process that keeps bone healthy and strong.

Metastatic cancer can amplify this process; it can affect the normal balance between new and old bone and change the structure and function of the bone. Osteoblastic metastases develop when cancer cells invade the bone and cause too many bone cells to form. The bone becomes very dense, or sclerotic. Osteoclastic metastases develop when metastatic cancer cells break down too much of the bone, making it very weak; in fact, holes may develop in the bones as the bone is destroyed.

In advanced prostate cancer, it is mainly the osteoclastic mechanism that destroys the bone.

From the results mentioned above (CT-scan), the emergency health care provider scheduled a follow-up with the on-call radio oncologist, and the follow-up took place on April 13, 2015. Following this appointment the radio oncologist concluded that no radiation treatment was necessary at this time. However, if the patient's headaches persist, the metastases would be handled with palliative radiotherapy.

The bone scan from March 30, 2015 confirmed what was seen in the current CT scan and demonstrates a disappearance as well as stability of some bone metastases. It also confirmed the appearance of new metastases (as mentioned earlier), some of which are at the skull level, making the radiologist confirm to a progressive worsening of known metastatic bone disease.

Due to the persistence of these symptoms of vertigo and migraines the patient saw his family doctor to get an MRI of the head and neck, which was performed on June 16, 2015 and showed no evidence of intracerebral metastasis. The MRI of the neck shown no evidence of bone metastasis to the cervical spine, but a discopathy was noted, especially at C5–C6 without evidence of spinal stenosis. As you will see below, the patient had palliative radiotherapy at the skull level that brought only minor improvements to his headaches and had no effect on his persistent vertigo. Therefore A is not the best answer.

Comment #1: Bone turnover biomarkers. The skeleton is the main location of prostate cancer metastases, affecting up to 90% of advanced castration-resistant prostate cancers [3]. Once in the bones, tumor cells disturb the equilibrium between bone formation by osteoblasts and bone resorption by osteoclasts [4]. As such, bone metastasis from prostate cancers is responsible for complications known as skeletal-related events (SREs). Examples of SREs include pathologic fractures, chronic bone pain, and spinal cord compression. These patients also present a decrease in quality of life, an increase in cost of treatment, and a decrease in survival odds versus those who have not had SREs [5]. All of the above information supports further investigation into bone biomarkers that could predict the risk of bone metastases and an increased severity of bone damage. This will be further discussed in the Case Report #9.

Evidence for the use of bone biomarkers for diagnosis and screening metastases progression remains unclear due to a lack of solid prospective studies. However, data is more optimistic for the use of bone biomarkers as an indicator of the response to treatment of bone metastases [6,7]. A superiority trial with Xgeva commonly used in the treatment of metastatic bone tumors in castration-resistant prostate cancer demonstrated a superior suppression of bone turnovers markers than zoledronic acid. For example, the median bone-specific alkaline phosphatase (AP) and N-telopeptide decreased significantly more compared to the baseline value in the denosumab group than in the zoledronic group [8].

Radium-223 (Ra-223; Xofigo) is another Health Canada-approved treatment for prostate cancer bone metastases and is included in the treatment plan for our patient either before or after chemotherapy, if necessary. In a phase II study prior to its approval, Ra-223 had significantly decreased or slowed down the increase of all the 5 bone biomarkers compared to the placebo group, including a median change of bone-specific AP of −65.6% with Ra-223 compared to a median change of 9.3% for the placebo group. These changes were associated to an average 3 weeks delay for a first SRE in the treatment group compared to placebo [8]. It seems that the monitoring of those bone turnover markers could help in the decision-making process for treatment of bone metastases [9]. Decrease in bone biomarkers could indicate successful treatment and support maintain the therapy while an increase in bone marker during a therapy could warrant a switch toward a more aggressive treatment alternative. This will be further discussed in the Case Report #9.

More specifically, bone-specific and nonspecific AP seems the most promising biomarker for the prognosis and response to treatments of bone metastasis. For our patient the AP level was very high at the time of diagnosis of his stage 4 prostate cancer, but it was reduced gradually to a normal level during his treatment with hormone therapy and Xgeva. Currently, his AP level is on the low side, and this is consistent with the fact that the metastatic activity is rather low and stable. This suggests that this biomarker is particularly relevant for assessing the severity of prostate cancer at the time of diagnosis [7]; however, it becomes less useful as a follow-up means to assess bone activity during treatment [7]. In fact a recent study found that prostate cancer cells express a tumor-derived AP and that this expression increases following bone metastasis. The expression of this tumor-derived form of AP was associated with tumor cell survival and increased tumor migration, while high levels of AP correlated with a decrease in the survival of patients suffering from prostate cancer with bone metastasis [10]. Therefore the fact that the AP level of our patient is currently rather low indicates the survival of our patient is favorable.

These observations could explain another recent finding that patients with fast ($\geq 5.42\,U/L/year$) alkaline phosphatase velocity (APV) have worse overall survival and bone metastasis-free survival than patients with slower AP kinetics [10]. In conclusion, bone biomarkers (alkaline phosphatase in particular) hold promising characteristics in terms of treatment follow-ups of bone metastases in advanced prostate cancer populations [11].

Comment #2: The decision to take Xgeva was relevant considering that medication inhibits bone resorption and permits the interruption of cancer-induced bone destruction. Fortunately, because of this mechanism of action, our patients noticed an important decrease in his bone pain with the use of Xgeva. To measure bone turnovers secondary to this medication, calcium levels were also used; however, his calcium level remains relatively stable during his treatment.

As briefly discussed earlier, many studies have already evaluated the prognostic value of bone turnover markers [7]. Among them, bone-specific AP, lactate dehydrogenase, and urinary N-telopeptide (Ntx) were associated with skeletal-related events, bone disease progression, and death in patients with solid tumors, including prostate cancer [1,2]. Similar results were observed by Coleman et al. in patients with prostate cancer treated with zoledronic acid. They found that high levels of Ntx were associated with a four- to sixfold increase in the risk of death [5]. On the other hand the normalization of the same bone biomarkers was associated with the reduced risks of skeletal complications [12], but Ntx has never been tested in our patients. Therefore we can easily assume that these biomarkers are relevant to our patient with bone metastatic prostate cancer treated with Xgeva.

(B) Benign positional vertigo. Benign positional vertigo (BPV) is the most common cause of vertigo. It causes a sudden sensation of spinning, like your head is spinning from the inside. If you have BPV, you can have brief periods of mild or intense dizziness [13]. An episode is generally triggered by changing the position of your head. In particular the following actions can trigger an episode of BPV: tilting your head up or down, lying down, or turning over and getting up. BPV can be annoying, but it is rarely serious except for when a person falls due to dizziness. BPV is the result of a disturbance inside your inner ear [13]. The fluid inside the tubes in your ear (semicircular canals) moves when your position changes; the semicircular canals are extremely sensitive.

BPV develops when small crystals of calcium carbonate that are normally in another area of the ear break free and find their way to the semicircular canal in your inner ear [13]. This causes your brain to receive confusing messages about your body's position. There are no major risk factors for BPV, but there is some indication that it could be an inherited condition.

Many diagnosed individuals have indicated that multiple relatives have also had the condition. Prior head injuries, osteoporosis, diabetes, or an inner ear condition can also make some people more prone to developing BPV.

The patient then consulted an ears, nose, and throat specialist (ENT) in early May 2015 for advice concerning the persistence of vertigo. The ENT did not believe in the possibility of labyrinthitis or in a reaction secondary to drugs or anxiety. Instead, he presented a diagnosis of BPV and recommended a follow-up physical therapy program for a vestibular rehabilitation.

The patient completed the series of vestibular exercises suggested by the physical therapist; after a few sessions, it was concluded that he did not have BPV. Furthermore, multiple canalith repositioning procedures performed by the physical therapist on this patient had no impact on his vertigo concluding in the absence of BPV.

On December 7, 2015 the patient consulted again the ENT specialist with the ENG results, which confirmed that our patient's persistent vertigo problems are not due to a problem in the inner ear, but rather a central problem without specifying the cause that further eliminated the diagnosis of BPV. Finally, considering the symptoms presented by the patient a BPV is unlikely. **(C) Disembarkment syndrome.** The physical therapist then suggested the presence of a syndrome that is similar to the "disembarkment syndrome," a rather rare and unlikely syndrome that usually occurs in people who have had long stays in a boat, which was not the case in our patient [13]. Although this term originally referred to the illusion of movement felt as an aftereffect of travel on water by ship or boat, some experts now include other types of travel (e.g., airplane, automobile, and train) as well as situations with novel movement patterns (e.g., reclining on a waterbed).

Most individuals experience this illusion of movement almost immediately after the cessation of the precipitating event, and the sensation usually resolves itself within 24 hours. This sensation is very common, and approximately 75% of all professional sailors experience it. However, for some individuals, this illusion of movement lasts for longer periods of time, as it did in our patient; in fact, it can last for weeks, months, and even years after the precipitating event. Persistent disembarkment syndrome has been defined as that which lasts longer than 1 month. Although a majority of the cases of persistent disembarkment syndrome resolve on their own within 1 year, there is possibilities that this syndrome persists over 12 months, as it did for our patient.

This persistent type of disembarkment syndrome was formally described in the medical literature in 1987 [13]. The reason the disembarkment syndrome becomes the persistent form of a few individuals (especially middle-aged women) and not in the vast majority of individuals is unknown. A leading explanation for the disembarkment syndrome is that the problem is not in the inner ear but rather in the brain.

This explanation is based upon studies that have demonstrated changes in the brain metabolism and functional brain connections of those individuals who have the disorder. Because of these changes the brain is able to adapt to an unfamiliar movement but it is unable to readapt once the movement has stopped. Although the reason for the persistence of this problem with readaptation is not completely understood, one theory suggests that certain movements (such as those experienced on a ship or boat) expose an individual to novel movement patterns in all planes of motion. During this time the brain must send signals to the body, so the muscles will be able to adapt to the novel movement patterns.

The most common symptoms associated with the disembarkment syndrome are rocking, swaying, and disequilibrium. Although this disorder may be accompanied by anxiety and depression, it is seldom accompanied by a true spinning vertigo. The symptoms of disembarkment syndrome usually feel worse when an individual is in an enclosed space or is attempting to be motionless, such as while lying down in bed. Stress and/or fatigue cause the symptoms to become more noticeable in some individuals. Although the symptoms often improve or even disappear during continuous movements, such as those experienced while driving a motor vehicle. Overall the disembarkment syndrome negatively affects an individual's quality of life, as in our patient.

Currently, there is no specific test to diagnose the disembarkment syndrome. For a diagnosis of disembarkment syndrome to be made the individual must subjectively report a history of a novel movement pattern (e.g., travel on water by ship or boat), the return to a normal environment, and the beginning of rocking, swaying, and disequilibrium sensations shortly thereafter. These symptoms begin immediately, not weeks or months later.

In order to rule out other causes of the symptoms, objective diagnostic procedures such as vestibular testing and radiological imaging may be performed. In individuals with the disembarkment syndrome, these examinations are usually normal. Currently, there is no single highly successful treatment approach for the disembarkment syndrome. Standard drugs prescribed for motion sickness (including meclizine and scopolamine patches) are usually ineffective in stopping or even decreasing the symptoms. Some treatments that have shown promise include vestibular rehabilitation, the use of benzodiazepines (such as valium), and the use of tricyclic antidepressants (such as amitriptyline). Unfortunately, these drugs had no impact in our patient.

Nevertheless, postural adjustment exercises, acupuncture, and others have helped the patient return gradually to full-time work. The symptoms of persistent vertigo and migraines have not disappeared with physical therapeutic exercises, but the patient has learned to live with them and was able to return gradually to work full time. Considering that our patient's persistent vertigo and migraine are completely different than the habitual presentation of the disembarkment syndrome, C is therefore unlikely.

(D) Paraneoplastic syndrome affecting the cerebellum. A paraneoplastic syndrome is a set of signs and symptoms that is the consequence of cancer in the body, but unlike in mass effect, it is not due to the local presence of cancer cells. In contrast, these phenomena are mediated by humoral factors (e.g., hormones or cytokines) excreted by tumor cells or by an immune response against the tumor [1,2].

Paraneoplastic syndromes are typical among middle-aged to older patients, and they most commonly present with cancers of the lung, breast, ovaries or lymphatic system (a lymphoma). Sometimes the symptoms of paraneoplastic syndromes show before the diagnosis of a malignancy, which has been hypothesized to relate to the disease pathogenesis.

In this circumstance, tumor cells express tissue-restricted antigens (e.g., neuronal proteins) triggering an antitumor immune response which may be partially or completely effective in suppressing tumor growth and symptoms [1,2], though the latter is rare. Patients present symptoms when this tumor immune response breaks immune tolerance and begins to attack the normal tissue expressing the (e.g., neuronal) protein.

On June 23, 2015 the patient then consults a neurologist, who believes in a paraneoplastic syndrome affecting the cerebellum. This syndrome takes the form of a cerebellar ataxia that can evolve in an impairment of fine motor skills and spontaneous falls. These people suffer from balance disorders, speech problems, coordination, movement, and writing disorders to various degrees that can appear at different times during their life and can become very disabling and worsen more or less quickly.

Because the patient is presenting mainly vertigo, yet no speech problems or writing disorders, a cerebellar ataxia is less likely. Just as the previous specialists, this neurologist assumes that the persistent vertigo in our patient would probably be secondary to the cancer itself or its treatment, and unfortunately no treatment is available. Thus our patient will live with his vertigo for the rest of his life unless his prostate cancer is better treated. It is very disappointing for the patient since his prostate cancer is incurable; therefore a paraneoplastic syndrome to explain our patient's persistent vertigo and migraine remains the most likely cause.

(E) Vestibular migraine. On October 7, 2015 our patient saw his neurologist for the follow-up of a lumbar puncture performed earlier. The results show no abnormalities in the cerebrospinal fluid. Combined with the MRI results the neurologist added the possibility of a vestibular migraine, a neurological disorder of the CNS resulting from a medical condition such as the cancer in our patient.

This vestibular migraine is often chronic and continuous with no specific trigger [13]. It is important to remember that the symptoms are always described as a disturbance in the balance, which means that it is essential to pay special attention to the risks of falls causing fractures [13]. This is particularly important for our patient with bone metastases at the pelvic and rib level. However, considering that our patient has no major balance disorders and that he never fell because of his persistent vertigo a vestibular migraine is less likely, but it is not far from the possible differential diagnosis list.

(F) Systemic inflammatory response syndrome (SIRS) is an inflammatory state affecting the whole body, frequently a response of the immune system to infection [14]. It is the body's response to an infectious or noninfectious insult. Although the definition of systemic inflammatory response syndrome (SIRS) refers to it as an "inflammatory" response, it actually has pro- and antiinflammatory components. SIRS is a serious condition related to systemic inflammation, organ dysfunction, and organ failure. It is a subset of the cytokine storm, in which there is an abnormal regulation of various cytokines. SIRS is also closely related to sepsis, in which patients satisfy criteria for SIRS and have a suspected or proven infection [14].

To rule out this possibility, in February 2017 the patient tried a loading dose of Decadron with tapering. Decadron is used for the treatment of a wide variety of diseases and conditions principally for its glucocorticoid effects as an antiinflammatory and immunosuppressant agent.

Interestingly, this N of 1 trial had successfully reduced the persistent headaches for a few weeks, but it had no impact on the persistent vertigo, which goes against this diagnosis. Even though our patient had a CRP of 73.6 ($N < 10$) and a sedimentation rate of 82 ($N = 15$) demonstrating a high level of inflammation secondary to this highly invasive prostate cancer, this is not sufficient enough to explain the persistent vertigo and migraine presented by our patient because the symptoms are also present, even though the CRP had returned to normal levels. Therefore this case presentation is not consistent with a SIRS, so F is not the best answer.

(G) Autoimmune syndrome. When an intruder invades the body, such as a cold virus or bacteria or a thorn that pricks your skin, the immune system works to protect it. It tries to identify, kill, and eliminate the invaders that might hurt you. But sometimes problems with the immune system cause it to mistake your body's own healthy cells as invaders and then repeatedly attack them. This is called an autoimmune disease. (Autoimmune means immunity against the self.) Your immune system is the network of cells and tissues throughout your body that work together to defend you from invasion and infection.

Your immune system has two parts: the acquired and the innate immune systems. The acquired (or adaptive) immune system develops as a person grows. It "remembers" invaders so that it can fight them if they come back [15]. When the immune system is working properly, foreign invaders provoke the body to activate immune cells against the invaders and to produce proteins called antibodies that attach to the invaders so that they can be recognized and destroyed. The more primitive innate (or inborn) immune system activates white blood cells to destroy invaders without using antibodies.

Autoimmune diseases refer to problems with the acquired immune system's reactions. In an autoimmune reaction, antibodies and immune cells target the body's own healthy tissues by mistake, signaling the body to attack them. Autoimmune diseases can affect almost any part of the body, including the heart, brain, nerves, muscles, skin, eyes, joints, lungs, kidneys, glands, digestive tract, and blood vessels. The classic sign of an autoimmune disease is inflammation, which can cause redness, heat, pain, and swelling [15].

How an autoimmune disease affects a patient depends on what part of the body is targeted. If the disease affects the joints, as in rheumatoid arthritis, the patient might have joint pain, stiffness, and loss of function. If it affects the thyroid, as in Graves' disease and thyroiditis, it might cause tiredness, weight gain, and muscle aches. If it attacks the skin, as it does in scleroderma/systemic sclerosis, vitiligo, and systemic lupus erythematosus (SLE), it can cause rashes, blisters, and color changes. Many autoimmune diseases don't restrict themselves to one part of the body. For example, SLE can affect the skin, joints, kidneys, heart, nerves, blood vessels, and more. No one is sure what causes autoimmune diseases; in most cases, a combination of factors is probably involved [15]. For example, you might have a genetic tendency to develop a disease and then under the right conditions an outside invader like a virus might trigger it [15].

To rule out this possibility the patient has received intravenous human immune globulins (Panzyga). The mechanism of action of IVIGs in the treatment of vertigo is not known. One possible mechanism may be the inhibition of the elimination of autoantibody-reacted cerebellum cells from the blood circulation by an IgG-induced Fc-receptor blockade of phagocytes. After having received the loading dose of immune globulins and during an additional 6 months of treatment the patient is still presenting his persistent vertigo and headaches, which goes against this diagnosis.

Unfortunately, this patient has a very low level of white cells, particularly in neutrophil levels, but after 2 months of Panzyga, this was already improved. This represents an important beneficial effect for the patient that justifies pursuing his treatment with Panzyga.

For the benefits of the reader, Panzyga is known to improve the immune system. For more information regarding this product, please consult the following link: http://www.octapharma.ca/fileadmin/user_upload/octapharma.ca/Product_Monographs/PANZYGA-PM-EN.pdf.

Discussion: It therefore becomes more and more obvious that the cancer itself or its treatment is primarily responsible for our patient persistent vertigo. It is also essential to point out that despite what the name may suggest, in many cases a vestibular migraine is not accompanied by headaches and can be of varying duration; that is, a shorter duration as in seconds or a longer duration as in several days to several weeks, which differentiates them from a typical migraine attack.

Because vestibular migraines may vary in different ways the diagnosis is often difficult to establish. This can often be treated with beta blockers and low-dose antidepressants. Because our patient's blood pressure is rather low the neurologist opts for an empirical treatment with 25 mg of Elavil an antidepressant, nortriptyline, an antidepressant, with a follow-up in 4 months. After 4 months, the Elavil was discontinued because it had no effect. Then the patient was put on Topiramate, a medication to treat epilepsy or to use as a migraine prophylaxis. After many months of Topiramate, there was no success and no improvements, so this medication was also discontinued.

Meanwhile the patient's hemoglobin developed a macrocytic anemia mainly secondary to hormone therapy. Also note that this kind of anemia can also be responsible for the persistent vertigo in this patient. After few months, the macrocytic anemia improved, but the vertigo persisted, thus eliminating this cause as being responsible for the patient's persistent vertigo. An assessment of vitamin B12 was also performed, as a deficit in B12 can also explain the macrocytic anemia and be responsible for his persistent vertigo. Unfortunately the B12 level came back as normal, as it did for vitamin B6 and folic acid. Therefore it became more and more obvious that the cancer itself or its treatment is primarily responsible for our patient's persistent vertigo and headaches.

REFERENCES

[1] Canadian Cancer Society. What is prostate cancer? Available at: http://www.cancer.ca/en/cancer-information/cancer-type/prostate/prostate-cancer/?region=on; 2017.

[2] Canadian Cancer Society's Advisory Committee on Cancer Statistics. 2017. Available at: http://www.cancer.ca/en/cancer-information/cancer-101/canadian-cancer-statistics-publication/?region=bc.

[3] Bubendorf L, Schopfer A, Wagner U, Sauter G, Moch H, Willi N, et al. Metastatic patterns of prostate cancer: an autopsy study of 1,589 patients. Hum Pathol 2000;31:578–83.

[4] Fohr B, Dunstan CR, Seibel MJ. Clinical review 165: markers of bone remodeling in metastatic bone disease. J Clin Endocrinol Metab 2003;88:5059–75.

[5] Coleman RE. Metastatic bone disease: clinical features, pathophysiology and treatment strategies. Cancer Treat Rev 2001;27:165–76.

[6] Brown JE, Sim S. Evolving role of bone biomarkers in castration-resistant prostate cancer. Neoplasia 2010;12:685–96.

[7] Archambault W, Plourde G. Biomarkers in advanced prostate cancer. Mini review. J Pharmacol Clin Res 2017;3:1–9.

[8] Lipton A, Fizazi K, Stopeck AT, Henry DH, Brown JE, Yardley DA, et al. Superiority of denosumab to zoledronic acid for prevention of skeletal-related events: a combined analysis of 3 pivotal, randomised, phase 3 trials. Eur J Cancer 2012;48:3082–92.

[9] Saad F, Lipton A, Cook R, Chen YM, Smith M, Coleman R. Pathologic fractures correlate with reduced survival in patients with malignant bone disease. Cancer 2007;110:1860–7.

[10] Hammerich KH, Donahue TF, Rosner IL, Cullen J, Kuo HC, Hurwitz L, et al. Alkaline phosphatase velocity predicts overall survival and bone metastasis in patients with castration-resistant prostate cancer. Urol Oncol 2017;35:460.e21–8.

[11] Rao SR, Snaith AE, Marino D, Cheng X, Lwin ST, Orriss IR, et al. Tumour-derived alkaline phosphatase regulates tumour growth, epithelial plasticity and disease-free survival in metastatic prostate cancer. Br J Cancer 2017;116:227–36.

[12] Smith MR, Cook RJ, Coleman R, et al. Predictors of skeletal complications in men with hormone-refractory metastatic prostate cancer. Urology 2007;70:315–9.

[13] Vestibular disorders association. Types of vestibular disorders. Is there more than one kind of vestibular disorder? Available at: https://vestibular.org/understanding-vestibular-disorder/types-vestibular-disorders; 2017.

[14] Kaplan LJ. Systemic inflammatory response syndrome. Medscape 2017. Available at: http://emedicine.medscape.com/article/168943-overview.

[15] Roddick J. Autoimmune disease. Types, causes, symptoms diagnosis and treatment. Healthline 2017;. Available at: http://www.healthline.com/health/autoimmune-disorders.

CHAPTER 6

Case Report #6—Bone Metastasis and Its Treatment

INTRODUCTION

In the meantime the patient gets the results of a bone scan completed on August 5, 2015. This bone scan, compared to the one carried out in March 2015, shows the persistence of metastatic lesions on the right scapula, the seventh rib, and the iliac bone. Among other things, there is a presence of metastatic lesions on the left pubic branch and the sacrum, bilaterally. This means that bone metastasis becomes a very important issue for our patient and he certainly wants to understand the best treatment modalities for managing this important condition, as he is aware that bone metastasis puts his life at higher risk of fatality and would certainly contribute to decreasing his quality of life. Now he is not much concerned about his prostate cancer by itself because his PSA is at an acceptable level, but he does consider bone metastasis as his No. 1 enemy.

Question #1: As this patient is still responding to hormone therapy and his No. 1 enemy is bone metastasis, what should be next best treatment(s) if the patient is still refusing chemotherapy?

(A) Xofigo
(B) Xgeva
(C) Radiation therapy
(D) Hormone therapy
(E) Zytiga
(F) All of the above
(G) A, B, C, and E are the best answers

Answer: G.

(A) Xofigo. Xofigo (radium Ra 223 dichloride) is a therapeutic alpha particle-emitting pharmaceutical with a targeted antitumor effect on bone metastases. Xofigo (radium Ra 223 dichloride) is indicated for the treatment of patients with castration-resistant prostate cancer with symptomatic bone metastases and no known visceral metastatic disease.

© 2018 Elsevier Inc.
All rights reserved.

Regarding the possibility of getting this medication, the patient met a specialist from the Montreal University Prostate Cancer Research Center in order to be involved in a research project using Xofigo. At the time of this appointment the research protocol was not yet ready, and the patient needed to wait for months before he could undertake this research project, as it was a double-blind randomized control clinical trial where neither the patient nor the physician knew what molecule the patient would receive. Being in the control group represented a risk to the patient, and so the patient decided not to participate in this study. Furthermore, because at the time of the appointment with this specialist the patient was still responding positively to hormone therapy, which is an exclusion criterion in this research project, he was excluded from participating in this phase III double-blind randomized clinical trial.

Finally, it was mutually agreed upon with his oncologist that for now, the patient would continue the treatment on hormone therapy. The plan they both discussed is as follows: treatment with hormone therapy until the patient becomes hormone resistant. The patient will initiate Zytiga as a second-line hormone therapy, and then if we noticed an insignificant response, chemotherapy with Docetaxel and finally Xofigo would then be initiated. For more information on the beneficial effects and the adverse drug reactions associated with Xofigo, the readers may wish to consult the following link: http://omr.bayer.ca/omr/online/xofigo-pm-en.pdf.

(B) Xgeva. Xgeva is an inhibitor of the ligand RANK (see Case Report #3). By preventing the interaction between the RANK and its ligand, Xgeva inhibits the formation, function, and survival of osteoclasts, which reduces the resorption and interrupts the bone destruction caused by cancer. In order to improve his chances to relieve bone pain, the patient decides to take Xgeva 120 mg subcutaneous every 6 months (from January 2016) and then every month (from September 2017) for its anticancer effect on bone metastases. This drug is also used in this patient for the treatment of osteoporosis, which is a side effect associated with hormone therapy. For more information about Xgeva, please consult the following link: https://www.amgen.ca/products/~/media/e06e33ed57d8457c8bc1509ceced41d9.ashx.

Comment #1: As mentioned earlier with Xgeva, the patient is more at risk for hypocalcemia and its consequences (an altered mental state, tetany, seizures and heart arrhythmia); therefore it is now essential for the patient to follow his blood calcium level rigorously. In addition, before and while taking Xgeva, the patient should make sure that his teeth are in a good condition and must therefore be regularly examined by a dentist to prevent a possible osteonecrosis of the jaw, if a dental surgery or invasive procedure is required. Since the patient started to take Xgeva, his blood calcium level remains normal. He also consulted his dentist before taking Xgeva to see if dental procedures were required. Then he was regularly followed by his dentist and no change in his teeth was noted.

(C) **Radiation therapy.** This therapy seems to have the best benefit/risk ratio for the treatment of pain associated with bone metastases and has lower side effects than chemotherapy (see Case Report #3). The patient met the radio oncologist on September 29, 2015, and was prescribed a single external radiotherapy treatment, which aims to curb the activity of bone metastases and reduce bone pain at the skull level and the pubic branches of the sacrum, bilaterally. The patient was treated on October 2, 2015 and noticed no significant side effects, except a light pain on radiation site, which was considered transitory and of little importance.

Despite the radiotherapy, pain in the pelvis (iliac crest bilaterally) and headaches remained but noticed some improvements in the radiated areas. The patient then met his oncologist on October 26, 2015, and new radiotherapy treatments were prescribed. The patient received radiotherapy treatment for the seventh rib on November 9, 2015 and November 11, 2015, he also received treatment for metastasis located on two ribs on the right side and the left iliac crests. Following the previous radiation treatments, the patient noticed a significant reduction in ribs and pelvic pains. On December 11, 2015, the patient saw his oncologist and mentioned that radiation treatments to date have had good results and that the most active bone metastases are under control. So, the oncologist recommended continuing the current treatment with hormone therapy and follow-up in 6 months with the radio-oncologist, as necessary. The patient got two more radiation therapy treatments and currently, his bone pain is under better control without pain medications.

(D) **Hormone therapy.** As explained above after 2½ years of IAS the patient's PSA level is still low (<0.2) but has increased in the last 2 months to the point (0.9) that it was justified to reintroduce Casodex, which was done on June 5, 2017. This means that the patient is no longer on IAS for a certain period of time. He may be allowed to return on IAS if his PSA level decreases to a low level and remains stable for at least 6 months on the combination Zoladex and Casodex. Then, Casodex can again be removed, as necessary. However, in September of 2017, his PSA increased to 2.1, therefore a second-line hormone therapy was introduced (see below; item E). This option is no longer suitable for our patient because he is now hormone resistant.

The rest of the blood work was normal including calcium, alkaline phosphatase, and LDH values, which are important biomarkers of bone remodeling. The mean platelet volume and hemoglobin were normal, but the white blood and neutrophil cells were low. Fortunately, as explained above, the treatment with Panzyga has corrected these two laboratory abnormalities.

(E) Zytiga. Abiraterone acetate (ZYTIGA) is converted in vivo to abiraterone, an androgen biosynthesis inhibitor. Specifically, abiraterone selectively inhibits the enzyme 17α-hydroxylase/C17, 20-lyase (CYP17), which is at the top of the steroidogenesis pathway (enzyme cascade). Because of that, it is more efficient than other drugs in the same category but acting at a lower level of the steroidogenesis enzyme cascade.

The above enzyme is expressed in and is required for androgen biosynthesis in testicular, adrenal, and prostatic tumor tissues. It catalyzes the conversion of pregnenolone and progesterone into testosterone precursors, DHEA, and androstenedione, respectively, by 17-α hydroxylation and cleavage of the C17, 20 bound. CYP17 inhibition also results in an increased mineralocorticoid production by the adrenals. Androgen-sensitive prostatic carcinoma responds to treatment that decreases androgen levels. Androgen deprivation therapies, such as treatment with GnRH agonists or orchiectomy, decrease androgen production in the testes but do not affect androgen production by the adrenals or in the tumor. For more information on the mechanism of action of Zytiga, please consult the following link: http://www.janssen.com/canada/sites/www_janssen_com_canada/files/product/pdf/zyt08252016cpm_snds_191178.pdf.

Many patients can stabilize their prostate cancer for several years on hormone therapy. However, as our patient has an aggressive and invasive cancer, it is unlikely that he will respond to this treatment over an extensive period of time. Eventually his cancer becomes resistant to the first-line hormone therapy, so the patient opted for a second-line hormone therapy with Zytiga on September 18, 2017. It was thus discovered that abiraterone acetate (Zytiga) combined with prednisone (a powerful antiinflammatory health product) taken orally can inhibit the production of the majority of the hormones that stimulate cancer cells of hormonal origin [1,2].

These hormones are produced by the adrenal glands as well as cancer cells, even in patients who have undergone a physical castration. Studies have shown that abiraterone acetate combined with prednisone could allow a significant improvement in the survival of subjects covered by this medication compared to those who are not. It has also been demonstrated that when combined with prednisone, abiraterone acetate effectively delayed the progression of cancer and prolonged the lives of patients who have not yet been submitted to chemotherapy [1,2]. In addition, possible side effects of this treatment are fewer than those associated with chemotherapy with Docetaxel or treatment with Xofigo.

More recently, new results from the STAMPEDE and LATITUDE trials presented at the American Society of Clinical Oncology (ASCO) 2017 Annual Meeting have propelled abiraterone acetate into first-line use (i.e., before chemotherapy) [1,2]. At present, the drug is approved for use after the failure of androgen deprivation therapy (ADT), but the new data shows substantial benefit when used earlier on, in combination with ADT. The findings confirm that in newly diagnosed disease, treatment with abiraterone acetate is largely displacing chemotherapy from the current treatment program [1,2]. As explained below, abiraterone acetate can be used in many situations before the use of chemotherapy when indicated to the patient on a case-by-case basis, but represents the next best option for our patient after the first-line hormone therapy. This drug is also known for its beneficial effects on the quality of life of patients suffering from prostate cancer, as a reduction in the time for opioid use suggest a reduction in pain or a better quality of life.

In the most recent STAMPEDE trial [1] a total of 1917 patients underwent randomization. The median age was 67 years, and the median PSA level was 53 ng per milliliter. A total of 52% of the patients had a metastatic disease, 20% had a node-positive or node-indeterminate nonmetastatic disease, and 28% had a node-negative, nonmetastatic disease; and 95% had a newly diagnosed disease. The median follow-up was 40 months. There were 184 deaths in the combination group as compared with 262 in the ADT-alone group (hazard ratio, 0.63; 95% confidence interval [CI], 0.52–0.76; $P < .001$); the hazard ratio was 0.75 in patients with nonmetastatic disease and 0.61 in those with metastatic disease [1]. There were 248 treatment-failure events in the combination group as compared with 535 in the ADT-alone group (hazard ratio, 0.29; 95% CI, 0.25–0.34; $P < .001$); the hazard ratio was 0.21 in patients with nonmetastatic disease and 0.31 in those with metastatic disease [1].

Results from the recent LATITUDE trial in nearly 1200 men show that after a median follow-up of 30.4 months, the overall median survival was significantly longer in the abiraterone acetate group than in the placebo group with a hazard ratio for the death of 0.62; (95% confidence interval [CI], 0.51–0.76; $P < .001$) [2]. The median length of radiographic progression-free survival was 33.0 months in the abiraterone group and 14.8 months in the placebo group (hazard ratio for disease progression or death, 0.47; 95% CI, 0.39–0.55; $P < .001$) [2]. Significantly better improvement for the time until pain progression and prostate specific antigen progression ($P < .001$) along with symptomatic skeletal events ($P = .009$). It is interesting to note that these findings led to the unanimous recommendation by the independent data and safety monitoring committee that the trial is unblinded and crossover was allowed for patients in the placebo group to receive abiraterone [2].

The benefit from the early use of abiraterone acetate observed in these studies is at least comparable to the benefit of docetaxel chemotherapy, which was observed in prior clinical trials, but abiraterone acetate is much easier to tolerate. These findings also indicated that the addition of abiraterone (plus prednisone) to ADT can potentially be considered a new standard of care for patients with high-risk, newly diagnosed metastatic prostate cancer [1,2]. However, several severe side effects were noted in the abiraterone group including high blood pressure (20% vs 10%), low potassium level (10.4% vs 1.3%), and liver enzyme abnormalities (5.5% vs 1.3%).

As discussed earlier, a few years ago the treatment program for prostate cancer was shaken up by the CHAARTED study [3], which showed a survival benefit associated with the addition of chemotherapy to hormone therapy in a patient sensitive to hormone therapy. These new data provide an important alternative to chemotherapy, namely the use of abiraterone acetate [1,2].

Although it is challenging to put the existing data for chemotherapy and abiraterone side by side, however, it appears that the benefits of survival mirrors or exceeds the benefit the authors have seen with chemotherapy [1,2]. Chemotherapy brought with it significant toxicities like nerve damage, fatigue, and decreases in blood counts that are often very difficult for patients with cancers to manage. Abiraterone brings the same or greater levels of effectiveness against prostate cancer with far fewer side effects [1,2]. The real benefit is that in addition to chemotherapy, in the highest-risk patients, we now have an oral option that may have less toxicity with similar long-term benefit in patients with newly diagnosed metastatic, castrate-sensitive prostate cancer. Therefore this represents a very interesting choice for our patient with a favorable benefit/risk ratio. Furthermore, this ratio seems superior to using chemotherapy at this time. For more information about Zytiga, please consult the following link: https://www.janssen.com/canada/sites/www_janssen_com_canada/files/product/pdf/zyt07222015cpmf_snds_177122.pdf.

Discussion: With the relative lack of effective cancer treatment options, research will continue to be a broad field of investment for pharmaceutical companies. The current number of products on the market and in development is representative of a fast-growing area and includes the following.

Hormone therapy includes agonists and antagonists of the LHRH and antiandrogens. These drugs are considered a form of chemical castration. Goserelin (Zoladex) and Leuprolide (Lupron, Eligard) are considered LHRH agonists. They decrease the rate of testosterone production by increasing drastically and quickly the production of testosterone which triggers a negative feedback to the level of the hypothalamus and blocks the release of LHRH. However, rising levels of testosterone (testosterone flare), although very short, is responsible for adverse effects, including pain in the bones and spine. Other agonists of the LHRH approved by Health Canada include: Histreline acetate (Vantas) and the Triptorelin pamoate (Trelstar) [4].

The most recent antagonist of the LHRH registered in Canada is degarelix acetate (Firmagon). Degarelix acetate is administered via subcutaneous injections. Its effectiveness as a testosterone suppressor is considered equivalent to leuprolide acetate, but it has a greater progression-free survival than leuprolide. In addition, degarelix acetate provides extended control of bone metastases.

As discussed above, abiraterone acetate (Zytiga) is an inhibitor of the enzyme CYP17 responsible for the synthesis of androgens in certain types of cells, including the prostate cells. Zytiga is generally administered orally daily to treat advanced and castration-resistant prostate cancer and is now recommended even in patients with castration-sensitive prostate cancer. Since Zytiga does not block testosterone synthesis at the testicular level, patients who have not responded to an orchiectomy or hormone therapy must combine this treatment with a true LHRH analogue. In addition, Zytiga is orally administered concomitantly with prednisone to avoid adverse effects related to the decrease in the synthesis of other no androgenic hormones by Zytiga.

In Canada, the top four antiandrogen drugs in clinics are bicalutamide (Casodex), cyproterone acetate (Cyproterone), flutamide (Euflex) and the enzulatamide (Xtandi). Bicalutamide (Casodex) is a nonsteroidal antiandrogen indicated in cases of metastatic prostate cancer usually administered as adjuvants following the removal of the testicles or chemical castration with an LHRH. As explained above, because the patient's PSA has increased recently, Casodex was reintroduced on 05 June 2017. Because the reintroduction of Casodex was unable to reduce the PSA level and even worse the PSA level still continue to rise, Zytiga plus prednisone was added on September 18, 2017 and Casodex was discontinued.

It seems that the effectiveness of antiandrogen drugs after castration is not very different between drugs. Casodex acts as an antagonist of the androgen receptors in the cytosol and blocks the action of testosterone. Adverse reactions associated with it include hot flashes, tiredness, a decrease in libido, and high blood pressure. More recently, severe cases of heart failure have been associated with Casodex. For more information about the mechanism of action of this product, its efficacy, and safety, consult the following link: https://www.astrazeneca.ca/content/dam/az-ca/downloads/productinformation/CASODEX%20-%20Product%20Monograph%20 2016-11-17.pdf.

Cyproterone acetate (Cyproterone) is a synthetic steroidal antiandrogen drug administered orally and via intramuscular injections. It prevents the growth of testosterone-dependent tumors by blocking androgen receptors. In addition, cyproterone acetate acts as an agonist of the progesterone at the pituitary gland level which decreases the release of LH. However, its antineoplastic activity is too weak to be used on its own. The adverse effects associated with it include hair loss, decreased libido and energy, weight gain, and an increase in stool [4–6]. For more information about the benefit/risk ratio of this product, please consult the following link: http://e-lactancia. org/media/papers/Cyproterone-PM-AAPharma2010.pdf.

Flutamide (Euflex) and its active metabolite 2-hydroxyflutamide are structurally similar to Casodex. Thus its effectiveness and toxicity profiles are similar to those of Casodex. Euflex and 2-hydroxyflutamide primarily block the interaction between dihydrotestosterone and the androgen receptors via the formation of inactive complexes, thus inhibiting the translocation of androgens inside the cell nucleus. Euflex is available in tablet form and prescribed in cases of metastatic and castration-resistant prostate cancer. However, for the latter, Euflex has a response rate of 15%, without benefit on survival [6,7]. For more information about the benefit/risk ratio of this product, please consult the following link: https://hemonc.org/docs/packageinsert/flutamide.pdf.

Enzalutamide (Xtandi) is a new type of androgen antagonist prescribed for men with a castration-resistant prostate cancer who have previously received or not received chemotherapy (docetaxel). Its mechanism of action is different from the other antiandrogens [8]. Indeed, it inhibits the action of androgens by blocking the signal transmitted by the androgen receptors and the interaction between them and the androgenic hormones [4–6]. Xtandi increases the patient's life expectancy by reducing the size of the tumors and their rate of growth and by lowering the PSA levels. In studies on Xtandi (administered orally) the most common adverse effects include headaches, musculoskeletal pain, and diarrhea [8] Because Xtandi is blocking androgen at a lower level in the steroid enzymatic cascade, it is considered slightly less efficient than Zytiga. For more information about the benefit/risk ratio of this product, please consult the following link: https://www.accessdata.fda.gov/drugsatfda_docs/label/2012/203415lbl.pdf.

When the prostate cancer is no longer responding to an orchiectomy, to analogues of LHRH, or to the androgen antagonists, patients must rely on third- and fourth-line alternatives. These treatments have poor results or associated with very important adverse effects. However, they are last resort options that are still worth a try. Small doses of corticosteroids, such as prednisone or dexamethasone, inhibit the production of androgens by adrenal glands. They produce a therapeutic response in approximately 33% of the cases of castration-resistant prostate cancer [9].

The use of estrogen analogues agonist has already been the standard in the treatment of prostate cancer, but it has been replaced by LHRH analogues and androgen antagonists. Since exogenous estrogen has higher gastrointestinal adverse effects, edema and blood clot risks this type of drug is no longer recommended. Furthermore, the response rate of these analogues varies between 20% and 40% [9].

Finally the antifungal medicine ketoconazole is a last option for advanced and metastatic prostate cancers. Ketaconazole blocks the production of several hormones through a mechanism similar to Zytiga. However, the antifungal also inhibits the production of cortisol. Thus ketoconazole should regularly be given with a corticosteroid to minimize significant adverse effects associated with low levels of cortisol [7–9]; however, its efficacy is rather poor.

Until recently, chemotherapy with docetaxel (taxotere) was considered to be the treatment of reference but only in the case of castration-resistant metastatic prostate cancer. However, the important phase III randomized controlled CHAARTED trial suggested that it would be more beneficial to add docetaxel to antiandrogen therapy for cases of castration-sensitive metastatic prostate cancer [3].

In a randomized sample of 790 patients the overall survival of those who received docetaxel plus antiandrogen therapy was 13.6 months higher than patients who only received antiandrogen therapy (57.6 months vs 44 months; RR = 0.61; $P < .001$). A second major study, STAMPEDE, which analyzed over 1000 patients, corroborated the results obtained by the CHAARTED study [3,10].

In the STAMPEDE study, docetaxel was coadministered with prednisone. The coadministration of corticosteroids reduces the incidence of neutropenia, an important adverse effect of the taxanes, compared to the administration of docetaxel alone [10]. Therefore in light of these data, six cycles of docetaxel and prednisone at the beginning of an antiandrogen therapy was recommended as the standard treatment for all cases of prostate cancer.

However, there is still not enough data to confirm its use in first-line prostate cancer therapy, and the data from the recent study on Zytiga (STAMPEDE and LATITUDE trials) suggest that the latter is better than docetaxel for this indication. Of course, readers who want to know more about Docetaxel, its efficacy, and its adverse effects can consult the Canadian Product Monograph for docetaxel by visiting the following link: http://products.sanofi.ca/en/taxotere.pdf.

Unfortunately, not all patients are responding favorably to docetaxel. However, a new form of chemotherapy called cabazitaxel (JEVTANA) is available and is likely to bring some hope to patients suffering from prostate cancer by extending their life [9]. JEVTANA belongs to the taxanes class.

JEVTANA is an antineoplastic agent that acts by disrupting the microtubules network in cells. JEVTANA binds to tubulin and promotes the assembly of tubulin into microtubules while simultaneously inhibiting their disassembly. This leads to the stabilization of microtubules, which results in the inhibition of mitotic and interphase cellular functions. Very common adverse reactions associated with Jevtana are anemia, leucopenia, neutropenia, thrombocytopenia, and peripheral neuropathy (including peripheral sensory and motor neuropathy). For more information about this product in terms of efficacy and safety, please consult the following link: http://products.sanofi.ca/en/jevtana.pdf.

REFERENCES

[1] James ND, de Bono JS, Spears MR, for the STAMPEDE investigators. Abiraterone for prostate cancer not previously treated with hormone therapy. N Engl J Med 2017;377:4.
[2] Fizazi K, Tran NP, Fein L, for the LATITUDE Investigators. Abiraterone plus prednisone in metastatic, castration-sensitive prostate cancer. N Engl J Med 2017;377:4.

[3] Sweeney CJ, Hui Chen Y, Carducci M, Liu G, Jarrard DF, et al. Chemohormonal therapy in metastatic hormone-sensitive prostate cancer. N Engl J Med 2015;373:737–46.

[4] Canadian Cancer Society. What is prostate cancer? Available at: http://www.cancer.ca/en/cancer-information/cancer-type/prostate/prostate-cancer/?region=on; 2017.

[5] Guirguis NC, Bissanda NK, Kamel MH. Counseling the patient with prostate cancer: 2016 update. Urol Nephrol Open Access J 2016;3:1–3.

[6] American Cancer Society. Hormone therapy for prostate cancer. Available at: https://www.cancer.org/cancer/prostate-cancer/treating/hormone-therapy.html; 2017.

[7] Gomella LG, Jaspreet S, Lalass C, Trabusi EJ. Hormone therapy in the management of prostate cancer: evidence-based approaches. Ther Adv Urol 2010;2:171–81.

[8] Graff JN, Alumkal JJ, JJ DCG, et al. Early evidence of anti-PD-1 activity in enzalutamide-resistant prostate cancer. Oncotarget 2016;7:52810–7.

[9] Prostate Cancer Canada. Prostate cancer. Available at: http://www.prostatecancer.ca/Prostate-Cancer/About-Prostate-Cancer/Prostate-Cancer; 2017.

[10] James ND, Sydes MR, Clarke NW, Mason MD, Dearnaley DP, et al. Addition of docetaxel, zoledronic acid, or both to first-line long-term hormone therapy in prostate cancer (STAMPEDE): Survival results from an adaptive, multiarm, multistage, platform randomised controlled trial. Lancet 2016;387:1163–77.

CHAPTER 7

Case Report #7—How to Improve Quality of Life

INTRODUCTION

In the previous case reports, we have discussed treatment modalities mainly in terms of efficacy and safety, benefit/risk ratio, and in terms of overall survival, but we have not really discussed the improvement of quality of life. Obviously, having a treatment that is more efficient and safer will contribute to increasing the quality of life as discussed previously. However, when a patient is asking you about improving the quality of life he is not necessarily thinking about only the pharmaceutical aspects of it.

Question #1: Among the following, what are the elements to consider in the improvement of quality of life without pharmaceutical aids?

(A) Quality of diet
(B) Regular physical activity
(C) Stress management
(D) Keeping intimacy alive
(E) All of the above

Answer: E.

(A) Quality of diet. An important way to improve your quality of life is to complement, not replace your medical treatment by better lifestyle habits that include a balanced diet, regular physical activity, and better stress management, as well as by keeping intimacy with a partner alive. Prostate cancer remains a leading cause of mortality in men, and the prevalence continues to rise worldwide, especially in countries where men consume a Western-style diet. Epidemiological, preclinical, and clinical studies suggest a potential role for dietary intake in the incidence and progression of prostate cancer. Low carbohydrate intake, increased intake of omega-3 fats, green tea, almond milk, tomatoes (for their content in lycopene, or supplements of lycopene 20–50 mg/day), flaxseeds, chia, and others have shown promise

Prostate Cancer
https://doi.org/10.1016/B978-0-12-815966-8.00007-2

© 2018 Elsevier Inc.
All rights reserved.

in reducing prostate cancer risk or progression. On the other hand, higher saturated fat intake and higher β-carotene consumption may increase the risk of prostate cancer. A U-shape relationship may exist between a prostate cancer risk and folate, vitamin C, vitamin D, and calcium [1]. Despite inconsistent and inconclusive findings the potential benefits of a balance diet for the prevention and treatment of prostate cancer is promising [1]. The combination of all the beneficial factors for prostate cancer risk reduction in a healthy dietary pattern may be the best dietary advice. This pattern includes high consumption of fruits and vegetables, reduced refined carbohydrates, total and saturated fats, and reduced cooked meats [1].

However, our patient is known to have a very well-balanced diet that is low in fat and added sugar (simple sugars); that is, a Mediterranean-type diet. He consumed mostly fish, little red meat (fewer than 500 g per week), no transformed or processed foods (rich in sugars and added sodium), and a large quantity of fruit and vegetables every day. We also noticed that during the hormone therapy, our patient gained a good amount of weight. This further supports the importance of following an appropriate balanced diet during treatment. Our patient was also advised to have a diet composed of a good variety of food in order to avoid loss of appetite that patients with cancer often face.

Comment #1: It is important to understand that it is not red meat itself that is damaging, but mainly the processing of the meat and the presence of environmental contaminants like pesticides, herbicides, fungicides, and others that contaminate the foods.

Comment #2: The patient has a master's degree in nutrition and is therefore aware of what an appropriate balanced diet is. He makes sure to have optimal nutrition that includes antioxidants (i.e., substances that help fight free radicals) such as blueberries, juice of wheat grass and other nutrients recognized as having positive effects on the treatment of cancer and its side effects. To learn more about eating well following a diagnosis of prostate cancer, access the Nutrition Guide for Men with Prostate Cancer at the following link: http://www.bccancer.bc.ca/nutrition-site/Documents/Patient%20Education/nutrition_guide_for_men_with_prostate_cancer.pdf.

(B) Physical activity. When analyzing the studies that make up this case report regarding the role of physical activity in prostate cancer, we noticed that physical activity has a positive effect on improving the quality of life of patients having this health problem. Positive results were observed in reducing fatigue and stress, improving sexual functioning, as well as improving the metabolic profile in these patients [2,3]. Our patient was very active physically and made round trips of more than 30 km by bicycle per day to go to his work. Currently, he cannot bike very much, but does walk more than 30 minutes twice a day with his dog. He added yoga and Tai-Chi at least once a week which is a good way to stay active and fight stress.

The patient also has a bachelor's degree in physical activity and is therefore well informed about the benefits of physical activity in patients with cancer. Physical activity and regular stretching activities are also effective means for the patient to manage the stress associated with his health status. Even though the patient has noticed a big decrease in his capacity of doing physical activity, he still wants to do some in order to reap its benefits and obviously in order to remain physically autonomous. This is particularly important for our patient as he observed a decrease in his ability to perform physical activities including walking. Obviously, it is of major importance to stay autonomous as long as possible. The patient also recommends staying active mentally either by continuing working or engaging in your own activities and hobbies including writing, painting, and others. This helps with the stress management of cancer and offers a good sense of accomplishment. It is not true that having cancer equals a burden for the society; these patients have still a lot to offer to the society.

(C) Stress management. A cancer diagnosis understandably brings about feelings of stress, fear, aggressiveness, and anxiety. However, uncontrolled or prolonged stress can affect a patient's outcome when it comes to curing the disease. Stress can negatively affect treatments being used for prostate cancer. The body responds to this pressure by releasing hormones such as epinephrine and norepinephrine that increase blood pressure, elevate heart rate, and raise blood sugar levels. There is also evidence that stress during cancer can shorten the length of telomeres, which are the ends of chromosomes that deteriorate with aging. There is no currently evidence that stress causes cancer, says the National Cancer Institute, but not controlling stress can affect the quality of life for cancer patients and reduce their overall survival [1].

There is also animal research showing that stress promotes tumor progression. It was found that stress may also negatively affect the response to the treatments being used in prostate cancer patients [4]. Researchers at Wake Forest School of Medicine examined mouse models with prostate cancer that had been subjected to stress. The team measured levels of apoptosis (cell cancer death) as a result of the medical treatment given to reduce tumors. When compared to the mice that were unstressed the stressed mice exhibited a significantly reduced response to the drug. Interestingly, chemically induced stress via injection with adrenaline (epinephrine) also blocked cancer cell death [4].

Our patient is already doing physical activity to manage his stress but decided to add meditation to his stress management arsenal. Occasionally, he gets massages to complement his stress management strategies. The results of studies suggest that a regular practice of meditation improves various symptoms suffered by patients or survivors of cancer such as their quality of life.

With meditation, our patient can more easily manage his cancer and its associated effects on the body and mind. By practicing meditation for about 20 minutes per day before bedtime, the patient noticed better rest and less pain and anxiety.

(D) Keeping intimacy alive. Keeping intimacy alive is a good way to maintain or increase your quality of life. Whether you have been with your partner for decades or just for a short period, or whether you are dating different people or are looking for a relationship, intimacy is likely to be an important part of your quality of life. Intimacy may involve sexual activity, or it may be more emotionally driven, but in either case, intimacy makes our lives richer.

As a man with advanced prostate cancer like our patient, you may be concerned about the potential effects of the disease on intimacy either physically or emotionally. Whether you are single or part of a committed couple, you have the right to take steps toward a fulfilling life of intimacy. In order to achieve this objective, good communication and open mind are essential.

It is important that you discuss with your health care provider how sexual functioning can be adversely affected by an advanced prostate cancer diagnosis. Many people feel uncomfortable discussing sexual functioning and related intimacy issues. As a man with prostate cancer, your sexuality may be strongly affected, as you feel like you are losing your masculinity. The feeling of not having full sexual ability can lead to a sense of anger, loneliness or even depression.

Maintaining open communication between you and your partner, you can keep your intimacy alive by allowing the both of you to navigate through the essential task of redefining your sexual relationship. Fortunately, there are potential strategies for dealing with sex and intimacy. Although sexual intimacy can be an uncomfortable subject to talk about, your physician, nurse or another member of your health care team may be able to provide you with relevant guidance, support groups or educational materials on this issue.

It is important to understand that sexuality may include making your partner feel good in ways that go beyond penile penetration and ejaculation.

Dealing with an advanced prostate cancer diagnosis, deciding on treatment modalities, and trying your best to carry on as a normal man can be very challenging. It is particularly important to allow your partner to work through these challenges with you. Obviously, your sexual life with your partner may have an impact on your entire relationship. You should be proactive and have an open mind to new ideas and accept advice from others, including a friend facing the same difficulties, a health care professional, a couples' counselor, or your partner.

Discussion: It is well known that a good diet may help to reduce the risks of developing prostate cancer. As this has already been discussed above (Case Report #4), the current case report mainly discusses the role of a healthy lifestyle in improving the quality of life during treatment. One should note that during treatment, adjusting the diet may help the patient manage treatment side effects, and eating well during treatment may lower the risk of secondary cancers, as well as improve the patient's overall health. When patients are being treated for cancer, it is highly important to eat right and get adequate food, even though it might be more difficult than ever to adhere to a balanced diet.

The body needs extra energy and nutrients to fight against cancer and to repair healthy cells that may have been damaged as a side effect of treatments, particularly when chemotherapy and radiation therapy are used. At the same time, many cancer treatments including chemotherapy may induce side effects that drain strength and disrupt appetite. Therefore we can make sure that your patients are getting all the essential nutrients, vitamins, and minerals they need.

For many patients, it is recommended that they meet with a specialized dietician to ensure that their diet is adequate. Thus for our patient and other patients in treatment for cancer, it is important to make all the possible efforts to focus on foods that permit the recovery, preservation, or restoration of lost and damaged tissue in order to help them with their treatment. However, as discussed above, gaining too much weight is also not acceptable.

There is no doubt that antioxidants are beneficial for health. The effects of foods containing antioxidants (e.g., fruit and vegetables) are obvious. However, the effects of antioxidants taken in isolation are conflicting because all antioxidants are not equal with respect to the different sites and stages of cancer development.

In 2012, the results of a Cochrane systematic review were quite disturbing. In this review, the authors report that an increased risk of mortality was associated with beta-carotene and possibly vitamin E and vitamin A but was not associated with the use of vitamin C or selenium. The current evidence does not support the use of antioxidant supplements in the general population or in patients with various diseases [5]. Finally, they concluded that if you eat from each main food group in appropriate amounts, there is no need for vitamin and mineral supplements.

The American Institute for Cancer Research recommends eating 2–3 meals with fruit, vegetables, whole grains, and beans [1]. The Canadian Food Guide [6] recommends planning for three normal meals per day and one to three snacks each day and for each meal with choices from each of the four food groups. Fruit and vegetables should cover half of the plate while the other half is shared between grain products (±50%) and high-protein foods (±50%). It is also recommended to eat seven servings of fruit and vegetables, six to seven servings of grain products, three servings of dairy products, and two to three servings of meat and alternatives.

Men with prostate cancer are recommended to accumulate at least 150 minutes of moderate to vigorous intensity aerobic physical activity per week, in bouts of 10 minutes or more. It is also beneficial to add muscle and bone strengthening activities using major muscle groups, at least 2 days per week. More physical activity provides greater health benefits but the main goal for patients with advanced prostate cancer is to remain physically active as much as possible in order to maintain their physical condition.

In men with prostate cancer regular physical activity can increase overall well-being and quality of life as well as reduce stress. It may also help to minimize the side effects of prostate cancer treatments, improve bone health, prevent or minimize loss of muscle mass, help to maintain a healthy body weight, and stay autonomous. Sometimes, it might be difficult to perform regular physical activities, but it is important to push yourself to do at least the minimal amount of physical activity in order to stay autonomous as long as possible. For more information, the readers may wish to consult the following link: http://www.csep.ca/CMFiles/Guidelines/CSEP_PAGuidelines_adults_en.pdf.

Meditation and yoga could also represent safe tools for alleviating the symptoms caused by cancer. With respect to our patient, he strongly believes that the addition of meditation and yoga in his life is the element that has had the greatest positive impact on his quality of life, well beyond food and physical activity, as he is used to eat adequately and exercise regularly. Similarly, by keeping is intimacy alive, it means that he is not alone in facing his incurable prostate disorder. It also helps in reducing his stress and certainly improves his quality of life.

REFERENCES

[1] The World Cancer Research Fund (WCRF)/American Institute for Cancer Research (AICR). Policy and action for cancer prevention food, nutrition, and physical activity: a global perspective' report from the World Cancer Research Fund. Available at: http://www.aim-digest.com/digest/members%20over%20yr/Policy%20aND%20 ACTION%20.pdf; 2009.

[2] Boin GL, Seemann T, de Carvalho Souza MS, et al. The benefits of physical activity in men with prostate cancer—a systematic review. J Phys Educ 2016;27:1–14. e2729.

[3] Canadian Society of Exercise Physiology (CSEP). Canadian physical activity guidelines. Canadian sedentary behaviour guidelines. Your plan to get active every day. Available at: http://www.csep.ca/cmfiles/guidelines/csep_guidelines_handbook.pdf; 2017.

[4] Prostate Cancer Center of Excellence, Wake Forest School of Medicine. Available at: http://www.wakehealth.edu/Research/Comprehensive-Cancer-Center/Prostate-Cancer-Center-of-Excellence/Prostate-Cancer-Center-of-Excellence.htm.

[5] Bjelakovic G, Nikolova D, Gluud LL, Simonetti RG, Gluud C. Antioxidant supplements for prevention of mortality in healthy participants and patients with various diseases. Cochrane Database Syst Rev 2012. Available at: http://www.cochrane.org/CD007176/ LIVER_antioxidant-supplements-for-prevention-of-mortality-in-healthy-participants-and-patients-with-various-diseases.

[6] Health Canada. Canada food guide. Available at: https://www.canada.ca/content/dam/ hc-sc/migration/hc-sc/fn-an/alt_formats/hpfb-dgpsa/pdf/food-guide-aliment/print_ eatwell_bienmang-eng.pdf.

CHAPTER 8

Case Report # 8—Treatment Modalities in Development

INTRODUCTION

One day our patient came to the office asking for other promising options in treatment modalities. For 3 years, our patient was treated with hormone therapy to which he eventually became resistant. He had recently started taking Zytiga combined with prednisone, a second-line hormone therapy. At the same time the patient was also considering a chemotherapy-based treatment (Docetaxel or Jetvana) followed by a radium 223-based therapy (Xofigo). However, our patient wanted to know the best treatment options for him after the above modalities failed. He learned from the medical literature that various treatment options such as immunotherapy, gene therapy, nanotherapy, and vaccines are very promising upcoming treatments of prostate cancer.

Question # 1: Among the following, what are the most promising treatment approaches?

(A) Immunotherapy (check-point inhibitors)
(B) Gene therapy
(C) Nanotechnology
(D) Vaccines
(E) Oncolytic virotherapy
(F) All of the above

Answer F.

(A) Immunotherapy. A reduction of immune system activity plays a prominent role in the development of cancer, including prostate cancer [1,2]. It was discovered that cancerous cells can escape T cell responses to tumor-related antigens by multiple mechanisms. The role of active immunotherapy in cancer is to stimulate the immune response against cancer cells [1,2]. Prostate cancer is not ignored by the immune system, as demonstrated by the presence of tumor-infiltrating lymphocytes (TIL) in the prostatic

Prostate Cancer
https://doi.org/10.1016/B978-0-12-815966-8.00008-4

© 2018 Elsevier Inc.
All rights reserved.

tissue affected by this cancer. This characteristic permits prostate cancer cells to become suitable for immunotherapy [1,2]. The presence of well-defined antigens, largely limited to prostate tissue, permits these cells to be targeted by the immune system without the risk of systemic autoimmune reactions [1–3]. Active immunotherapy in prostate cancer can be conducted using multiple strategies such as dendritic cells, whole-cell vaccines, viral vectors, and DNA-based and peptide-based agents, as well as immune-stimulatory agents [1,2]. The only FDA-authorized immunotherapy for prostate cancer is the dendritic cell-based vaccine named Sipuleucel-T, which has an advantage in overall survival (OS), but not in progression-free survival [1–3]. This product was not authorized in Canada or Europe [2].

One of the main questions that is particularly relevant is this: Why do humans become resistant to androgen-deprivation therapy as the first-line treatment of prostate cancer. Eventually, most patients progress to a condition known as hormone-resistant or castration-resistant prostate cancer (CRPC), which is characterized by a lack of response to ADT [2,4]. Although new androgen receptor signaling inhibitors and chemotherapeutic agents have been introduced to overcome a resistance to ADT, patients still become resistant to these agents [2,4]. The majority of treatments that are currently used in clinical practice to treat patients with advanced prostate cancer modulate the androgen receptor signaling pathway [2]. However, the increased heterogeneity and genomic variability of advanced prostate cancer cells are frequently responsible for the resistance to treatment [5]. Many pathways, both androgen receptor-related and nonandrogen receptor-related, may play a role in the progression of hormone resistance [1–3]. In many instances, androgen receptor-independent pathways are involved in the growth of aggressive tumor cells with a high metastatic potential. Many nonandrogen receptor-directed drugs have been tested in hormone-resistant prostate cancer, mostly as monotherapy, and these studies provide conflicting results [1–3]. As discussed in recent literature the predictive biomarkers of responses to treatment are therefore needed to increase the success of future trials, as well as to properly select patients that would most likely benefit from these nonandrogen-related targeted therapies [6]. In addition, studies evaluating the combination of androgen receptor signaling modulators or chemotherapy with molecules that target androgen-independent pathways may provide new treatment options for advanced prostate cancer [4]; see Case Report #9.

Another highly pertinent question asked by the patient after having been explained some mechanism for resistance to hormone therapy is why we should choose immunotherapy. As discussed above, immunotherapy modulates and reinforce the patient's own immune responses against cancer in general and most probably against advanced prostate cancer [1–3]. Neoplastic cells naturally can escape from the control of the immune system, and the main goal of immune therapy is to get this short- and long-term control back [2]. Previous studies in advanced melanoma and lung cancer suggest a great potential of immunotherapy as an approach for other tumors' treatment including prostate cancer [1–3] with less side effects than chemotherapy. As mentioned earlier, prostate cancer was the first neoplasm in which a specific vaccine was discovered. Furthermore, there is a strong potential for synergistic combinations of immunotherapy with conventional cancer treatments including prostate cancer [7–10].

As briefly discussed above, recent evidence demonstrated that prostate cancer generates a variety of tumor-associated antigens (TAAs), including PSA, prostatic acid phosphatase, and prostatic-specific membrane antigens, which are capable of producing a clinical response through immunogenicity [1–3,7–10]. In fact the immune system is the major player in the control of disease, and cancer is not an exception. Immunotherapy is to boost the immune function and educate the immune response toward the tumor antigens [2].

Comment #1: Checkpoint inhibitors: The mechanism of action of these biological agents involves the inhibition of immunosuppressive drugs regulated by T cell signals and is a very important alternative in the treatment of many cancers [1–3]. However, these products are not working very well for the treatment of advanced prostate cancer, therefore they are not recommended as monotherapy for our patient at this time. Recent clinical trials with anti-CTLA-4 did not show statistical significance for prostate cancer. With anti-PD-1, there seemed to be inconsistent responses, but nothing seems to reach the level of what was seen in other tumor types [11]; however, the reasons for this are still unknown. More research is done to elucidate this poor response to prostate cancer, but this approach is very useful for other types of cancer, as specified below.

Comment #2: CTLA-4 inhibitors: Yervoy (ipilimumab) was the first agent in this class approved in Canada in 2012. The latter exercises its anti-neoplastic activity through the inhibition of the surface molecule CTLA-4 (cytotoxic T cell antigen #4). CTLA-4 is an inhibitory T cell protein mainly involved in the regulation of these cells. As this mechanism is relevant to any immune response, ipilimumab is potentially effective for the treatment of all cancer types; that is why it is currently under study in patients with prostate, lung, kidney, breast, and several other cancers [12].

The OS in a phase III clinical trial assessing metastatic or unresectable melanoma patients previously treated with standard treatments was 24% at 2 years against 14% in the control group. Death has been reported with this treatment, with serious side effects including colitis, hepatitis, toxic epidermal necrolysis, and the inflammation of the nerves and hormonal glands [12]. For more information about the benefits and side effects associated with this product, please visit the following link: http://www.yervoy.com/. As mentioned, this class of product is in development for many indications. Although associated with many side effects and sporadic responses in prostate cancer, it is still essential to follow the development of these molecules.

Comment #3: PD receptor antagonists: The PD-1 receptor (programmed death protein #1) is essential in the regulation of the peripheral T cells activity. Its endogenous ligands, PD-L1 and PD-L2 (programmed death protein ligands #1 and #2), are expressed by activated immune cells and tumor cells [13]. The literature confirms that some tumors escape the immune system by expressing PD-L1 and PD-L2 on their surface, allowing them to inhibit the infiltration lymphocytes (TIL) present via the PD-1 activation on the surface of immune cells [13].

The first monoclonal antibody anti-1 human IgG4 antibody Nivolumab is licensed under the name of Opdivo for the treatment of inoperable or metastatic melanoma and metastatic squamous NSCLC [14]. Unlike ipilimumab the efficacy and toxicity associated with the nivolumab does not seem to depend on the dose [14,15]. Opdivo showed a survival rate of 42% at 1 year against 24% for docetaxel. In addition, Opdivo has a higher median survival of 9.2 months compared to 6 months for docetaxel. In addition, adverse events including hematological disorders and nonhematological toxicity related to treatment occurred less frequently with Opdivo (58%) than docetaxel (86%) [14].

Comment #4: It thus seems that this molecule has a significant advantage over conventional chemotherapy with docetaxel in the treatment of the NSCLC, melanomas, and kidney cancers [14]. It certainly is a molecule that should attract readers by offering an acceptable compromise to chemotherapy with docetaxel. Again, although not authorized for the treatment of prostate cancer, this molecule certainly deserved to be followed in the near future. For more information about the benefits and side effects associated with this product, please visit the following link: http://www.opdivo.com/.

Comment #5: Keytruda (pembrolizumab) is another human IgG4 monoclonal antibody targeting PD-1. It currently appears as a second-line therapy for the treatment of unresectable or metastatic melanomas expressing the V600 BRAF gene mutation (personalized medicine) and for metastatic

NSCLC. In patients with melanomas that demonstrate a progression of the disease during the treatment with ipilumumab, Keytruda has a treatment overall rate of response of 24%. Among the patients treated, 86% showed a continuous response ranging from 1.4 to 8.5 months. This molecule is certainly of interest for the treatment of melanoma and lung cancers. For more information, the reader may consult the product monograph that can be found by visiting the following link: http://www.merck.ca/English/Products/Pages/Prescription-Products.aspx. Research is underway for this type of molecule and used in the treatment of other types of cancer. Although, research is promising in the treatment of prostate cancer using immune therapy, the patient will have to wait few years before knowing whether these treatment modalities will become available for him [2].

Comment #6: Strategies that use checkpoint inhibitors targeting PD-1/PD-L1 have been successful in treating other malignancies, but unfortunately were not successful in the treatment of prostate cancer [8,11,13]. We can hypothesis that the resistance to checkpoint inhibition can be overcome in prostate cancer through the targeting of myeloid derived suppressor cells (MDSCs), which are involved in suppressing T cell responses, which protects the cancer from the immune mediated action resulting in cancer cell destruction [8,11,13]. Importantly MDSCs are prevalent in metastatic prostate cancers. However, neither checkpoint inhibition nor MDSC targeting provides significant effect in mouse models of prostate cancer. However, the therapeutic effect is synergistic when combined and may be more efficient [8], but this still need to be studied.

Comment #7: We can suggest that a combined therapeutic approach may provide a more effective immunotherapeutic action in patients with advanced prostate cancer. In fact, it was recently discovered that the combination of two immunotherapy drugs could help increase the response of the immune system against advanced prostate cancer cells. However, as mentioned above a third brake called VISTA was found on a human tumor and may inhibit immune response [16]. Recently, in March 2017, Gao et al. [16] performed a clinical trial in patients with advanced prostate cancer combining two immunotherapy drugs targeting specific brakes on the immune system. This initial study, published in *Nature Medicine*, has explored whether they could increase immune cell infiltration of tumors by combining the LHRH agonist drug leuprolide with two rounds of the checkpoint inhibitor ipilimumab before surgery in patients with locally advanced prostate cancer. It is important to note that the genomic and immune analysis of the surgically removed tumors demonstrated high levels of penetration of the tumors by activated T cells. This study also indicated increased levels of immune-suppressing PD-L1 and VISTA. PD-L1 is known to interact with the immune checkpoint PD-1 on T cells, thus activating PD-1 to block the T cell. Gao et al. mentioned that driving T cells into the tumors would be the first step, but the next step would be to block PD-L1 and VISTA [8,16].

These results have been applied in an immunotherapy combination clinical trial that use ipilimumab to bring T cells into the tumor and the PD-1 inhibitor nivolumab to attack the PD-L1/PD-1 response [7]. The trial has enrolled 90 patients at 9 centers nationally. Although six therapies have been approved to treat patients with metastatic, hormone-resistant prostate cancer, none has provided a durable response [2]. The article published in *Nature Medicine* strongly suggest the importance of studying immune response longitudinally and that simply observing it at one point in time does not reflect what is occurring because the immune system changes so quickly [16].

Question #A2: True or False? Then the patients asks that if we combine immunotherapy with other cancer treatment modalities whether or not our chance of success in treating his advanced prostate cancer would increase.

Response: True.

Antiandrogen therapy is the main therapy for prostate cancer, and agents targeting the androgen receptor pathway continue to be developed such as Zytiga (abiraterone), which is a second-line hormone therapy that the patient is currently taking. Because antiandrogen therapy has immunostimulatory effects as well as direct antitumor effects, androgen receptor-targeted therapies could be combined with other anticancer therapies, including immunotherapies. A recent study was performed to evaluate whether an antigen-specific mechanism of resistance to antiandrogen therapy [9] may result in enhanced androgen receptor-specific T cell immune recognition, and whether this might be strategically combined with an antitumor vaccine targeting the androgen receptor. Antiandrogen therapy increased the androgen receptor expression in human and murine prostate tumor cells in vitro and in vivo [9], and this increased expression persisted over time. The increased androgen receptor expression was associated with recognition and cytolytic activity by androgen receptor-specific T cells. Furthermore, antiandrogen therapy combined with vaccination, specifically a DNA vaccine encoding the ligand-binding domain of the androgen receptor, led to improved antitumor responses as measured by tumor volumes and delays in the emergence of CRPC tumors in two murine prostate cancer models (Myc-CaP and prostate-specific PTEN-deficient mice) [8]. Taken together, this data suggests that antiandrogen therapy combined with androgen receptor-directed immunotherapy targets a major mechanism of resistance, the overexpression of the androgen receptor.

Several completed and ongoing studies have shown that the combination of cancer vaccines or checkpoint inhibitors with different immunotherapeutic agents, hormonal therapy (enzalutamide), radiotherapy (radium 223), DNA-damaging agents (olaparib), or chemotherapy (docetaxel) can enhance immune responses and induce more dramatic, long-lasting clinical responses without significant toxicity [2,8]. The goal of prostate cancer immunotherapy does not have to be the complete eradication of the advanced disease, but rather the return to an immunologic equilibrium with an indolent disease state [9,10]. In addition to determining the optimal combination of treatment regimens, efforts are also ongoing to discover biomarkers of immune response. With these concerted efforts, the use of immunotherapy in the treatment of prostate cancer seems more promising [2,8].

(B) Gene therapy. Cancer is a disease caused by mutations and or epigenetic changes in tumor suppressor genes and oncogenes that populate the host genome [5]. It is well established that most of the genetic events in cancer result from a series of accumulated, acquired genetic lesions [5,17]. These gene alterations involve structural changes such as mutations, insertions, deletions, amplifications, fusions, and translocations, or functional changes such as heritable changes without changes in nucleotide sequence [5,17–19].

However, very often no single genomic change is found, and multiple changes are commonly found in most cancers including prostate cancer. Approaches to cancer gene therapy include three main strategies: the insertion of a normal gene into cancer cells to replace a mutated (or otherwise altered) gene, genetic modification to block a mutated gene, and genetic approaches to directly kill the cancer cells [17–19].

Gene therapy is a method that attempts to introduce genetic material including DNA or RNA in a cell in order to alter the abnormal gene expression resulting from a genetic anomaly often associated with significant disease including cancer [17–19]. Primarily at the experimental stage, the gene therapy currently has many clinical trials of which the majority is for the treatment of cancer. As it is not simply enough to insert the appropriate genetic material in a diseased cell, vectors (i.e., means of transport) can be used in order to ease the transport of the genetic changes inside the cell that is to be rectified. These vectors are classified into two categories: the viral and nonviral vectors [17–19].

Briefly, all viruses (e.g., adenoviruses, retroviruses, and others) bind to their hosts and introduce their genetic material into the host cell that could be cancer cells as part of their replication cycle [5]. This genetic material contains basic "instructions" for how to produce more copies of these viruses, hacking the body's normal production machinery to serve the needs of the virus. The cancer cell will carry out these instructions and produce additional copies of the virus, leading to more and more cells becoming infected. Some types of viruses insert their genome into the host's cytoplasm, but do not actually enter the cell. Others penetrate the cell membrane disguised as protein molecules and enter the cell [5].

On the other hand, nonviral methods present certain advantages over viral methods with their simple large-scale production and low host immunogenicity. Previously, low levels of transfection and expression of the gene held nonviral methods at a disadvantage; however, recent advances in vector technology have yielded molecules and techniques with transfection efficiency similar to those of viruses [5].

The prostate is an especially suitable target for gene therapy, as it is expendable after the reproductive years and is easily accessible through a transrectal, transperitoneal, or transurethral approach [17–19]. Furthermore, using the regulatory sequences of prostate-specific proteins, such as PSA, prostatic acid phosphatase, prostate-specific membrane antigen, probasin, and other human glandular kallikreins to drive the expression of therapeutic genes, therapy can potentially be targeted to prostate cancer cells throughout the body using a systemic approach [17–19].

Numerous potentially therapeutic genes and delivery systems are currently being developed and evaluated. Overall the different gene therapy modalities can be grouped into three main categories: immunomodulatory, corrective, and cytoreductive.

- Immunomodulation, simply stated, attempts to augment the body's immune response to improve the immune system's natural ability to seek out and destroy cancer cells.
- Corrective gene therapy, which is already being used by many investigators in the treatment of patients with prostate cancer, involves the replacement or inactivation of a defective gene, such as a mutated tumor suppressor gene, or a dominant oncogene that has been found to play a role in the pathogenesis or progression of prostate cancer.
- The replacement of a defective *p53* tumor suppressor gene is an example of the type of corrective gene therapy that has produced the greatest interest.

The FDA has very recently approved Kymriah (tisagenlecleucel) for certain pediatric and young adult patients with a form of acute lymphoblastic leukemia (ALL). For more information, the readers may wish to consult the following link: https://www.fda.gov/downloads/BiologicsBloodVaccines/CellularGeneTherapyProducts/ApprovedProducts/UCM573941.pdf.

Although this approach is very promising in the treatment of prostate cancer the patient will still have to wait a few years before this treatment becomes available for him or for other patients suffering from prostate cancer. Fig. 1 provides an explanation on the gene therapy mechanism of action.

Comment #1: With the gene therapy approach, different vectors both viral and nonviral are employed. The six most frequently used viral vectors include those derived from adenovirus, retrovirus, poxvirus, adeno-associated virus, herpes simplex virus, and lentivirus, while the nonviral approaches includes calcium based molecular compounds, lipofection, and direct injection of naked DNA or RNA [17–19]. Furthermore, gene therapy vectors can be engineered to produce a variety of therapeutic proteins, and researchers are investigating the safety and effectiveness of different types of vectors in hundreds of clinical trials in the United States. For more information, please consult the following website: https://clinicaltrials.gov/.

One of the most commonly used vectors in gene therapy is the adenovirus vector (Fig. 1). While adenovirus vectors are very efficient at delivering genes, adenovirus vectors can cause toxic effects that limit the amount of

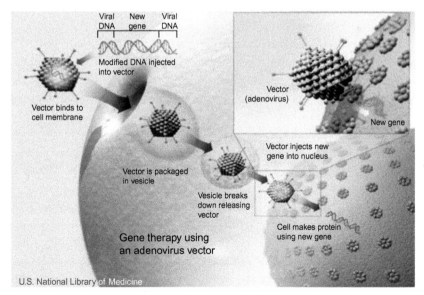

Fig. 1 Gene therapy. A new gene is injected into an adenovirus vector, which is used to introduce the modified DNA into a human cell. If the treatment is successful, the new gene will make a functional protein that corresponds to the inserted gene.

vectors that clinicians can give to patients. New adenovirus gene therapy vectors are currently tested in animals before human clinical trials begin. It is important for both researchers and regulatory agencies including the FDA, the EMA, and Health Canada to know how well these animal studies can predict safety in humans [5].

This new knowledge will enable researchers to design safer and more effective gene therapy vectors for treating prostate cancer. The objective is to improve the safety and efficacy of adenovirus and other vectors, especially when administered through the vascular system [5]. Researchers are studying additional novel mediators and pathways that control innate immune responses, as well as how this contributes to toxicity caused by adenovirus and other vectors. Understanding these mediators and pathways is an essential step toward our goal of developing safer vectors and new ways to limit vector-induced toxicity [5].

These are only few examples to demonstrate how the field of gene therapy is progressing in the treatment of prostate cancer. By looking at the following website: https://clinicaltrials.gov/, the readers will find that there >50 studies that are completed but not published, that others are in phases I, II, III in their development or are currently recruiting suggesting a great interest for gene therapy in the treatment of advanced prostate cancer. However, before gene therapy becomes available to our patient, further development is required. We should expect that this therapy would be available for our patients in the next few years. But patients can have access to this therapy through the Special Access Program or by being involved in clinical studies.

Comment #2: Chimeric antigen receptor (CAR) therapy: With the FDA approval of the first gene therapy (Kymriah) in the United States, we are entering a new era in medical innovation with the ability to reprogram a patient's own cells to attack a cancer (personalized medicine). Kymriah is a CD19-directed genetically modified autologous T cell immunotherapy indicated for the treatment of patients up to 25 years of age with B cell precursor ALL that is refractory to chemotherapy. The concept is that by engineering the T cells to recognize antigens on the surface of the tumors, we can bypass the normal antigen presentation by peptide and HLA Class I/II and engage T cells (or NK cells) to kill the tumor targets.

Kymriah is a personalized treatment that uses a patient's own T cells. In CAR T therapy a patient's T cells are extracted and cryogenically frozen. The cells are then genetically altered to have a new gene that codes for a protein called a chimeric antigen receptor (CAR). This protein directs the T cells to target and kill ALL cells with a specific antigen on their surface. The genetically modified cells are then infused back into the patient.

In a pivotal clinical trial of 63 children and young adults with ALL, 83% of patients that received the CAR T therapy had their cancer go into remission within 3 months. At 6 months, 89% of patients who received the therapy were still living, and at 12 months, 79% had survived. These results confirmed that this treatment is really efficient in humans.

CAR T therapy is also associated with life-threatening side effects in some patients, including neurological toxicity and cytokine release syndrome. Furthermore, another company ended a CAR T study earlier this year after patients died from cerebral edema. However, no patients treated with Kymriah have died from that complication. For more information on Kymriah, consult the following FDA approved package insert: https://www.fda.gov/downloads/BiologicsBloodVaccines/CellularGeneTherapyProducts/ApprovedProducts/UCM573941.pdf. The approval of Kymria has lead the International Society for Cellular Therapy (ISCT) to confirm that this approval increased enthusiasm in the field of cell therapy by boosting investment in products at all stages of drug development. For more information on ISCT visit the following link: http://www.celltherapysociety.org/. This approval creates also optimism in clinicians and researchers working on cancer, including prostate cancer.

Comment #3: Cancer stem cells (CSCs): Another potential strategy for aggressive prostate cancer is to target cancer stem cells (CSCs) [20–22], that can originate in any cell type of a particular tumor. CSC is believed to be a major factor contributing to the resistance to radiotherapy, conventional chemotherapies, most probably hormone therapy and for the development of metastasis. Therapeutic strategies aimed at targeting specific surface markers of CSCs, the key signaling pathways in the maintenance of self-renewal capacity of CSCs, such as the ATP-binding cassette transporters that mediate the drug-resistance of CSCs, dysregulated microRNAs expression profiles in CSCs, and immunotherapeutic strategies developed against precancerous stem cells (PCSCs) surface markers are also promising [20–22].

In a recent study [22] using a human prostate cancer cell line (PC3) the authors targeted the epithelial cell adhesion molecule (EpCAM), a protein that is overexpressed in CSCs from breast, colon, pancreas, and prostate cancers. The results demonstrated that higher levels of EpCAM correlated with the proliferation and metastasis potential of prostate cancer cells [5].

To produce CSCs-specific T cells the researchers isolated human peripheral blood cells (PBLs) and enriched them in T cells. Using retroviruses, these researchers genetically engineered T cells to have an EpCAM-specific

CAR. These new T cells were able to kill human prostate cancer cell lines in culture (ex vivo) [22].

Furthermore, when mice were injected with tumor cells, followed by administration of T cells bearing EpCAM-specific CAR, tumors did not metastasize. In contrast, mice that did not receive these modified cells had tumor metastasis after 27 days. In addition, treated mice showed prolonged survival, as animals treated with CAR-expressing PBLs were alive 80 days after treatment, whereas only 1/3 of nontreated mice survived. These data indicate that PBLs targeting EpCAM can have significant antitumor effects in prostate cancer [22].

The authors concluded that despite the low expression of EpCAM on PC3 tumor cells, EpCAM-specific PBLs had significant antitumor activity against PC3, probably by targeting the CSCs of prostate cancer. This data suggests that the adoptive transfer of T cells targeting CSC antigens is a promising therapeutic approach for treating prostate cancer [5,22].

(C) Nanotechnology. The emergence of nanotechnology has drawn much attention recently for applications in medicine such as "nanomedicine." Nanomedicine combines engineering, physics, biology, chemistry, mathematics, and medicine and strives to improve disease detection, imaging, and drug delivery through the use of nanodevices [23]. Nanotechnology includes the materials, devices, and delivery systems for disease diagnostics, prevention, and treatment.

Both researchers and the pharmaceutical industry have shown particular interest in nanotechnology for medical applications with potential benefits for patients. Nanotechnology-based medicine can bypass both physiological and biological barriers, such as the blood-brain barrier, endothelial barriers, cell membranes, and even nuclear envelopes, achieving both passive and active disease targeting [23]. Nanosized particles are normally composed of thousands of atoms and exhibit unique physical and biochemical properties with a high surface area for therapeutic loads such as cancer drugs. However, challenges to nanomedicine exist, such as safety issues, bulk manufacturing issues, and compatibility with the human body of nanosized drug delivery systems. Meeting these challenges is essential for the successful application of drug-loaded nanoparticles in the field of pharmaceutics.

Recent advancements in cancer nanotechnology facilitate the diagnosis and provide therapy for prostate cancer. Nanotechnology has the potential to fight tumors on the location, where the cancer begins. It is well known that there is a need to improve the therapeutic availability and the effectiveness of conventional chemotherapeutic agents for prostate cancer. To do so, many therapeutic natural products have been developed with nanotechnology that can specifically target and deliver a variety of agents including chemotherapy drugs to destruct the prostate cancer cells without causing any damage to the healthy cells [23,24].

Currently, theranostic natural products (NPs) have been developed specifically to target the prostate cancer cells using targeting ligands and to release the anticancer agents in a controlled and time-dependent manner for cancer therapy in combination with assisted imaging to monitor the effectiveness of the therapy in real time [23]. The natural products and surface-modified polymers and metallic natural products have evolved as promising nanomaterials for targeted prostate cancer treatment. Again, although, research is promising in the treatment of prostate cancer using this approach, the patient will have to wait some years before the nanotechnology treatment of prostate cancer becomes available for him. Fig. 2 presents an example of the use of nanotechnology using liposomes for drug delivery.

Liposome for drug delivery

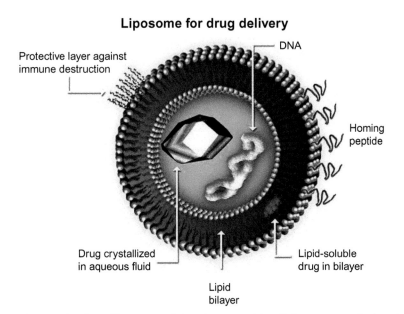

Protective layer against immune destruction

DNA

Homing peptide

Drug crystallized in aqueous fluid

Lipid-soluble drug in bilayer

Lipid bilayer

Fig. 2 Nanotechnology. This approach has the potential to fight tumors at the location where the cancer begins. To do so, many natural therapeutic products have been developed with nanotechnology that can specifically target and deliver a variety of agents including chemotherapy drugs to destruct the cancer cells without causing any damage to the healthy cells and its structure.

(D) Vaccines. There are several cell types that recognize and destroy tumor cells, such as macrophages, antigen presenting cells (APCs), CD8$^+$ (cytotoxic) T cells, cytotoxic T lymphocytes cells (CTLs), and natural killer (NK) cells. The presentation of tumor antigens to T cells by APCs including monocytes and dendritic cells (DCs) is required to stimulate adaptive immunity. The proliferation and generation of antitumor effects are stimulated by the antigen-specific T cells through cytokines and direct killing by CTL. However, tumor cells are developing pathways to suppress immune responses and escape from the immune system resulting in clinical progression. Most investigated mechanisms of escape from the immune system, include the modulation of immune-inhibitory (checkpoint) pathways to suppress T cell activity, and the disruption of antigen processing and presentation [2,25].

Tumors can also recruit and promote the developing immunosuppressive cells, such as regulatory T cells (Tregs) and myeloid-derived suppressor cells (MDSCs). Additionally the release of immunosuppressive factors, such as transforming growth factor-β, interleukin-6, and interleukin-10, might be directly or indirectly mediated by tumors [7–9]. These factors contribute to the development of the immunosuppressive microenvironment for the tumor. The cancer immunotherapies currently under investigation can be classified into four strategies for cancer treatment according to their target.

The first strategy is designed to augment the frequency of T cells in a patient specific to one or more tumor-associated antigens. The second strategy is a T-cell checkpoint blockade, such as CTLA-4, PD-1, or PD-L1 (see above). The third strategy is the use of T cells engineered to express a chimeric antigen receptor (CAR). The last strategy of immunotherapy is to disrupt or otherwise modify the immunosuppressive tumor microenvironment [8,11].

The four main types of vaccine-based immunotherapy investigated in prostate cancer can be classified as autologous, cell-based, viral-based, or peptide-based vaccines [26]. Prostate cancer is a disease for which cancer vaccines have shown survival benefits, with sipuleucel-T being the first cancer vaccine to receive FDA approval for the treatment of many malignancies. This autologous DC vaccine consists of prostatic acid phosphatase (PAP) and granulocyte-macrophage colony-stimulating factor (GM-CSF). Siluleucel-T provided the first solid evidence that cancer vaccines could provide a real benefit in clinical outcomes for patients with prostate cancer [25].

Side effects from the vaccine tend to be milder than those from hormone therapy or chemotherapy. Common side effects can include fever, chills, fatigue, back and joint pain, nausea, and headache. They most often start during the cell infusions and last for no more than a couple of days. A few men may have more severe symptoms, including problems breathing and high blood pressure, which usually get better after treatment. For more information about sipuleucel-T (Provenge), please consult the following link: https://www.fda.gov/downloads/BiologicsBloodVaccines/CellularGeneTherapyProducts/ApprovedProducts/UCM210031.pdf.

Other vaccines are currently in development for the treatment of prostate cancer. One of them is GVAX that is in early clinical investigation. This vaccine is a cell-based vaccine derived from LNCaP and PC3 both human prostate cancer cell lines genetically modified to secrete GM-CSF [25]. The results of phase 1/2 trials for patients with metastatic prostate cancer demonstrated that the PSA level decreased and stabilized, while a median OS time of 35.0 months with high-dose treatment was observed [26].

A second promising vaccine for prostate cancer is PROSTVAC (PSA-TRICOM), which is a PSA-targeted, poxvirus-based vaccine consisting of a heterologous prime boost (vaccinia or fowlpox virus vector) and three costimulatory molecules (TRICOM; B7.1, ICAM-1 and LFA-3) serving to increase the PSA-specific immune response [26]. Tumor-specific CTL responses and prolonged OS in castration resistant prostate cancer patients have been observed in a multicenter phase 2 clinical trial of PROSTVAC. Subsequently a phase 3 randomized, placebo-controlled trial of PROSTVAC (NCT01322490) is currently enrolling 1200 patients with metastatic castration-resistant prostate cancer (CRPC). Patients are randomized to receive PROSTVAC with GM-CSF, PROSTVAC with placebo GM-CSF or double placebo, with OS as the primary endpoint [26].

Another new concept of vaccine for prostate cancer is the personalized peptide vaccine (PPV). In this "personalized" cancer vaccine strategy, appropriate peptide antigens for vaccination are screened based on preexisting host immunity, and up to four peptides are selected from a list of vaccine candidates in each patient [25,26]. Previous phase 1 studies of PPV for CRPC showed that PPV was well tolerated; decreases in the PSA levels in some patients have been reported, while other studies are currently in investigation. This means that vaccines may represent interesting treatment modalities for our patient. Unfortunately, none of the vaccines is currently available at the present time [2].

(E) Oncolytic virotherapy. Oncolytic viruses (OVs) are viral strains that can infect and kill malignant cells while sparing their normal counterparts. OVs can access cells through binding to receptors on their surface or through fusion with the plasma membrane and establish a lytic cycle in tumors while leaving normal tissue essentially intact [25,26]. Multiple viruses have been investigated in humans in the past century. For example, IMLYGIC (T-VEC/Talimogene Laherparepvec), a genetically engineered herpes simplex virus, is the first OV authorized for use in the United States and the European Union for patients with locally advanced or nonresectable melanoma.

Although OVs have a favorable toxicity profile and are impressively active anticancer agents in vitro and in vivo the majority of OVs have limited clinical efficacy as a single agent [25,26]. While a virus-induced antitumor immune response can enhance oncolysis, when OVs are used systemically, the antiviral immune response can prevent the virus reaching the tumor tissue and having a therapeutic effect [25,26]. Intratumoral administration can provide direct access to tumor tissue and is beneficial in reducing side effects.

Immune checkpoint stimulation in tumor tissue has been noted after OV therapy and can be a natural response to viral-induced oncolysis. Also, for immune checkpoint inhibition to be effective in treating cancer, an immune response to tumor neoantigens and an inflamed tumor microenvironment are required, both of which treatment with an OV may provide. Therefore direct and indirect mechanisms of tumor killing provide a rationale for clinical trials investigating the combination of OVs with other forms of cancer therapy including immune checkpoint inhibition [25,26]. As for the other types of therapy, the oncolytic virotherapy is far from being applicable in patients with prostate cancer. Fig. 3 provides a simple demonstration on the functioning of the oncolytic virotherapy.

Discussion: Immunotherapy continues to provide new and promising tools for fighting cancer, and scientists are exploring these molecules for multiple cancers including prostate cancer [27]. However, numerous challenges remain before it becomes applicable in human patients with advanced prostate cancer. Fortunately, immunotherapy is giving to the scientific community new options for patients, and some are becoming promising treatments [28]. There has been amazing progress in the last 10 years, but we're still in the very early days in the development of this approach. Again, before this treatment becomes available in patients with advanced prostate cancer, much more research needs to be done; currently, these therapies are far from being a cure [2]. At this moment, we are still trying to identify the patients for which these available therapies will be sufficient for long-term control of their cancer, but scientists are still trying to understand what it is about

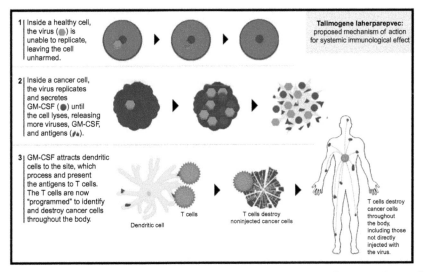

Fig. 3 Oncolytic virotherapy. Oncolytic viruses (OVs) are viral strains that can infect and kill malignant cells while sparing their normal counterparts. OVs can access cells via binding to receptors on their surface or through fusion with the plasma membrane and therefore establish a lytic cycle in tumors, while leaving normal tissue essentially intact.

the patients and their tumors that make the drugs efficient while others do not respond at all [8]. As mentioned, these therapies are considered less toxic than many types of chemotherapy because they involve the body's own immune system to fight the cancer [3,8]. The significant success of immune checkpoint inhibitors in shrinking advanced melanomas led to intensified efforts to apply the therapy to a variety of other cancers, including prostate cancer [28]. In 2016, the FDA, the EMA and Health Canada have approved immunotherapies for advanced forms of lung, kidney, bladder, head, and neck cancer, as well as Hodgkin lymphoma; this resulted in extending the survival for some groups of patients with the above cancers. The pace of approvals and new research has continued this year. Recently, the FDA, EMA, and Health Canada granted accelerated marketing authorization to the immunotherapy pembrolizumab, a programmed cell death protein 1 (PD-1) inhibitor, for patients with solid tumors [3,8]. The therapy is approved for adult and pediatric patients whose cancer has progressed despite prior treatment and who have no other treatment options. It is the first time that the FDA, Health Canada, and the EMA have approved a cancer treatment based solely on the presence of a genetic feature (BRAF V600 mutant) in a tumor; this indicates a new step toward personalized medicine in the treatment of cancer [8]. Pembrolizumab also was shown in a clinical trial to extend survival by 2 months for patients with nonsmall-cell lung cancer (NSCLC) compared with standard chemotherapy [29]. Within that group of patients, those with higher levels of PD-L1 had a median survival of 15 months compared with 8 months in the control group. Another PD-1 checkpoint inhibitor atezolizumab (Tecentriq) was the first such inhibitor to be authorized for some patients with advanced bladder cancer, as well as for patients with previously treated NSCLC [30].

Some of the biggest challenges in the field of immunotherapy lie in determining how and why patients develop resistance, as well as which specific groups respond to the therapy and why. The majority of immunotherapies fall in the 25%–30% patient response range [8]. In addition to immune checkpoint inhibitors, another major strategy is the T cell-based therapy, specifically chimeric antigen receptor (CAR) T cells, a form of adoptive cell transfer [8]. The approach involves drawing blood from patients and separating out the T cells. Using a killed virus the T cells then are genetically engineered to produce CARs on their surface. These receptors allow the T cells to recognize and attach to an antigen on the tumor cells. They then are infused back into the patient for the recognition and destruction of the cancer cells. CAR T cell therapy has been shown to be effective in patients with advanced blood cancers. (For more information about the CAR T cell, please look at the recent letter to editor from Dr. Plourde on this issue [5].) However, it still is unclear whether this method will be successful in patients with solid tumors. As discussed above the efficacy of immunotherapy has been more difficult to demonstrate in prostate tumors than in other types. Tumors could be explained by the fact that a lot of prostate cancer involves the bone, as in our patient, which makes some scientists question if there is something in the bone structure that makes immunotherapy more challenging [3]. Unfortunately, we currently cannot reassure our patient about the use of immunotherapy for the treatment of his stage IV prostate cancer. However, a large number of studies are currently ongoing in response to this question, but patients will have to wait at least a few years before it becomes available to them. It is possible that this treatment becomes available as monotherapy, but it most probably will be in combination therapy [2].

A rapidly emerging new immunotherapy approach seems promising; it is called adoptive cell transfer (ACT) and works by collecting and using patients' own immune cells to treat their cancer. There are several types of ACT (see ACT: TILs, TCRs, and CARs). But the one that is currently the most promising is called CAR T-cell therapy and was approved in August

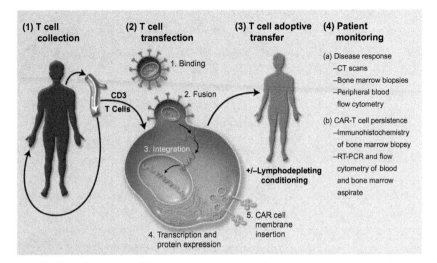

Fig. 4 CAR T cells. CAR T cells work by collecting and using the patient's own immune cells to treat their cancer, as explained above.

2017 for the treatment of patients up to 25 years of age with B cell precursor ALL [31]. Fig. 4 describes the process of CAR T cell functioning.

The FDA has authorized Kymriah based on the results of the pivotal open-label, multicenter, single-arm Phase II ELIANA trial [31], which was the first pediatric global CAR T cell therapy registration trial examining patients in 25 centers in the United States, EU, Canada, Australia, and Japan.

In this study, 68 patients were infused and 63 were evaluable for efficacy. Results from this Phase II ELIANA study show 52/63 (83%) (95% confidence interval [CI]: 71%–91%) of patients who received treatment with Kymriah achieved complete remission (CR) or CR with incomplete blood count recovery (CRi) within 3 months of infusion. In addition, no minimal residual disease (MRD); a blood biomarker that indicates a potential relapse was detected among responding patients. Median duration of remission was not reached (95% CI: 7.5-NE) [31].

The most common (>20%) adverse reactions in the ELIANA trial are cytokine release syndrome (CRS), hypogammaglobulinemia, infections-pathogen unspecified, pyrexia, decreased appetite, headache, encephalopathy, hypotension, bleeding episodes, tachycardia, nausea, diarrhea, vomiting, viral infectious disorders, hypoxia, fatigue, acute kidney injury, and delirium [31].

In the study, 49% of patients treated with Kymriah experienced grade 3 or 4 CRS, an on-target effect of the treatment that may occur when the engineered cells become activated in the patient's body. CRS was managed globally using prior site education and implementation of the CRS treatment algorithm. Within 8 weeks of treatment, 18% of patients experienced grade 3 or 4 neurologic events. There were no incidents of cerebral edema; the most common neurologic events were encephalopathy (34%), headaches (37%), delirium (21%), anxiety (13%), and tremor (9%).

Nevertheless, researchers caution that, in many respects, it is still early for CAR T cells and other forms of ACT, for questioning about whether they will ever be effective against solid tumors like breast, colorectal, and prostate cancer. However, in just the last few years, progress with CAR T cells and other ACT approaches has greatly accelerated, with researchers developing a better understanding of how these therapies work in patients and translating that knowledge into improvements in how they are developed and tested.

Therefore we should expect to see dramatic progress within the next few years; this can certainly be of interest to our patient, even though he is suffering from prostate cancer. For more information about this new authorized product, the readers are invited to consult the following link: https://www.fda.gov/downloads/BiologicsBloodVaccines/CellularGeneTherapyProducts/ApprovedProducts/UCM573941.pdf.

Recent developments have increased enthusiasm among cancer researchers; many now use therapeutic approaches in genetic manipulation to improve cancer regression and find a potential cure for the disease [5]. As stated, these therapies include transferring genetic material into a host cell through viral (or bacterial) and nonviral vectors, immunomodulation of tumor cells or the host immune system, and manipulation of the tumor microenvironment to reduce tumor vasculature or to increase tumor antigenicity for better recognition by the host immune system [5].

It is anticipated that gene therapy will play an important role in future cancer therapy as part of a multimodality treatment in combination with or following other forms of cancer therapy such as hormonal therapy, as with our patient. Gene therapy is already determined based on an individual's genomic constituents, as well as tumor specificities, genetics, and host immune status, to design a multimodality treatment that is personalized to each individual's specific needs. Finally, for our patient, we can be reassured that gene therapy will represent a very interesting option for him when he becomes resistant to the second-line hormone therapy. Gene therapy is very efficient, but it is still a concern from a safety perspective. Even considering the latter aspect, gene therapy already has a favorable benefit/risk ratio.

Of course, the genetic component is a nonmodifiable risk factor that increases the risk of developing cancer for many individuals. Gene therapy is, however, a treatment filled with hopes and promises for the future of so-called personalized therapy, which is a therapy adapted to the patient and its genetic components. This therapy is relatively new in the clinical world of cancer, but with the scientific interest for gene therapy in modern medicine, this predicts an increase of its use for the treatment of cancer including prostate cancer in the near future [5].

Nanotechnology research aims to create a method to identify a prostate cancer-specific drug; that is, a genotoxin to avoid drug resistance. The researchers seek to develop a biodegradable and biocompatible nanoparticle capable of targeting prostate cancer cells. It is hoped that nanotechnology and gene therapy studies will lead to a more personalized medical approach for patients with prostate cancer [5].

REFERENCES

[1] Janiczek M, Szylberg L, Kasperska A, et al. Immunotherapy as a promising treatment for prostate cancer: a systematic review. J Immunol Res 2017;4861570. 6 pages.

[2] Plourde G. Immunotherapy in advanced prostate. J Pharmacol Clin Res 2017;4(2):555635.

[3] Bilusic M, Madan RA, Gulley JL. Immunotherapy of prostate cancer: facts and hopes. Clin Cancer Res 2017;23(22):1–7.

[4] Cattrinia C, Zanardia E, Vallomea G, et al. Targeting androgen-independent pathways: new chances for patients with prostate cancer? Crit Rev Oncol Hematol 2017;118:42–53.

[5] Plourde G. Gene therapy in advanced prostate cancer: letter to the editor. J Pharmacol Clin Res 2017;4(1):555627.

[6] Archambault W, Plourde G. Biomarkers in advanced prostate cancer. Mini review. J Pharmacol Clin Res 2017;3(5):1–9.

[7] Lu X, Horner JW, Paul E, et al. Effective combinatorial immunotherapy for castration-resistant prostate cancer. Nature 2017;543:728–32.

[8] Cancer scope. Cancer 2017;123(21).

[9] Gill MD, Agarwal N. Cancer immunotherapy: a paradigm shift in the treatment of advanced urologic cancers. Urol Oncol 2017;35:676–7.

[10] Wei XX, Ko EC, Ryan CJ. Treatment strategies in low-volume metastatic castration-resistant prostate cancer. Curr Opin Urol 2017;27:596–603.

[11] Buchbinder EL, Anupam Desai A. CTLA-4 and PD-1 pathways, similarities, differences, and implications of their inhibition. Review article. Am J Clin Oncol 2016;39:98–106.

[12] Fellner C. Ipilimumab (yervoy) prolongs survival in advanced melanoma: serious side effects and a hefty price tag may limit its use. P T 2012;37:503–30.

[13] Topalian SL, Hodi FS, Brahmer JR, et al. Safety, activity, and immune correlates of anti–PD-1 antibody in cancer. N Engl J Med 2012;366:2443–54.

[14] Borghaei H, Paz-Ares L, Spigel M, et al. Nivolumab versus docetaxel in advanced nonsquamous non–small-cell lung cancer. N Engl J Med 2015;373:1627–39.

[15] Wolchok JD, Neyns B, Linette G, et al. Ipilimumab monotherapy in patients with pretreated advanced melanoma: a randomised, double-blind, multicentre, phase 2, dose-ranging study. Lancet Oncol 2010;11:155–64.

[16] Gao J, Ward JF, Pettaway CA, et al. VISTA is an inhibitory immune checkpoint that is increased after ipilimumab therapy in patients with prostate cancer. Nat Med 2017;23(5):551–5.

[17] Annan AC, Fisher PB, Dent P, Siegal GP, David T. Gene therapy in the treatment of human cancer. In: Coleman WB, Tsongalis GJ, editors. The molecular basis of human cancer. 2016. p. 811–41.

[18] Amer MH. Gene therapy for cancer: present status and future perspective. Mol Cell Ther 2014;2(27):1–19.

[19] Macpherson JL, Rasko JE. Clinical potential of gene therapy: towards meeting the demand. Intern Med J 2014;44(3):224–33.

[20] Yun E-J, Lo U-G, Hsieh J-T. The evolving landscape of prostate cancer stem cell: therapeutic implications and future challenges. Asian J Urol 2016;3:203–10.

[21] Maitland NJ, Collins AT. Prostate cancer stem cells: a new target for therapy. J Clin Oncol 2008;26(17):2862–70.

[22] Deng Z, Yanhong W, Ma W, Zhang S, Zhang Y-Q. Adoptive T-cell therapy of prostate cancer targeting the cancer stem cell antigen EpCAM. BMC Immunol 2015;16:1.

[23] Lakshmanan VK. Therapeutic efficacy of nanomedicines for prostate cancer: an update. Investig Clin Urol 2016;57:21–9.

[24] American Cancer Society. What's new in prostate cancer research? Available at: https://www.cancer.org/cancer/prostate-cancer/about/new-research.html; 2017.

[25] Geary MS, Salem AK. Prostate cancer vaccines; update on clinical development. Oncoimmunology 2013;2:e24523.

[26] Sweeney K, Halldén G. Oncolytic adenovirus-mediated therapy for prostate cancer. Oncolytic Virother 2016;(5)45–57.

[27] Olson BM, Gamat M, Seliski J, et al. Prostate cancer cells express more androgen receptor (AR) following androgen deprivation, improving recognition by AR-specific T cells. Cancer Immunol Res 2017;5(12):1074–85.

[28] Botta GP, Granowicz E, Costantini C. Advances on immunotherapy in genitourinary and renal cell carcinoma. Transl Cancer Res 2017;6(1):17–29.

[29] Keytruda: pembrolizumab: product monograph. Available at: https://pdf.hres.ca/dpd_pm/00040232.PDF; 2017.

[30] Tecentriq (atezolizumab): product monograph. Available at: http://www.rochecanada.com/content/dam/roche_canada/en_CA/documents/Research/ClinicalTrialsForms/Products/ConsumerInformation/MonographsandPublicAdvisories/Tecentriq/Tecentriq_PM_CIE.pdf; 2017.

[31] Buechner J, Grupp SA, Maude SL, et al. In: Global registration trial of efficacy and safety of CTL019 in pediatric and young adult patients with relapsed/refractory acute lymphoblastic leukemia: update to the interim analysis. European Hematology Association annual meeting. Abstract S476. Presented June 24; 2017.

CHAPTER 9

Case Report #9—Biomarkers of Prostate Cancer

INTRODUCTION

As our patient is a physician, he is highly concerned with his treatments and is following the medical literature in search of new and better biomarkers for the screening, diagnosis, and follow-up for his response to treatment. He came to the office asking about the role of biomarkers in the diagnosis and for the follow-up of his advanced prostate cancer's treatment. He wanted to discuss the new predictors of response and resistance to various treatments discussed in Case Report #8.

Question #1: Beside prostate-specific antigen (PSA), which of the following are the biomarkers that our patient should have used to improve the early detection of his prostate cancer?

(A) PSA derivatives
(B) PSA isoforms
(C) Prostate health index
(D) 4Kscore
(E) PC antigen 3 and Progensa
(F) C-reactive protein (CRP)
(G) None of the above was considered useful
(H) A and F were considered useful

Answer: H.

(A) PSA derivatives. At the moment the PSA blood test is the most widely used biomarker for prostate cancer screening and is considered superior to other risk factors, such as race and family history of prostate cancer for the screening [1]. However, the PSA blood test is not the optimal diagnostic biomarker; in fact, several groups agree that PSA has limitations in terms of specificity and sensitivity [2]. It does not seem to help reduce cancer mortality, and both the US Preventive Task Force and the Canadian Task

© 2018 Elsevier Inc.
All rights reserved.
93

Force on Preventive Health Care recommend against systematic screening for prostate cancer using PSA [3,4].

Considering the poor sensitivity and specificity of PSA levels in the screening of prostate cancer the development of more sensitive and more specific biomarkers allowing for early detection and diagnosis of prostate cancer is necessary. Similarly, for assessing treatment efficacy and treatment response, as well as for surveillance and detection of metastases, the development of new biomarkers becomes relevant [5].

In an effort to increase the sensitivity and the specificity of PSA the medical community have looked in different ways to interpret PSA blood levels. The derivative measures proposed include PSA kinetics such as: PSA doubling time, PSA velocity, PSA density, and age-specific PSA [6]. Unfortunately, scientific evidence to date concerning clinical advantages of those derivatives is not very optimistic [4].

None of these PSA kinetics biomarkers has demonstrated any clinically relevant additional value versus or in combination with the regular PSA analysis [7,8]. Even considering these controversies, we used the doubling time to make the decision of reintroducing Casodex for our patient. For >2½ years, our patient had a PSA <0.2 under treatment with Zoladex, then the PSA rose to 0.9 in the following 2 months for a PSA doubling time of around 1 month. For a patient already under treatment, such a doubling time was sufficient to reintroduce Casodex (see previous cases discussing the use of Casodex). During the following 2 months the PSA level has increased to 2.1, for a doubling time of around 1 month. Considering this situation, it was decided that a second-line hormone therapy (Zytiga+prednisone) was necessary. After the first months of Zytiga+prednisone the PSA remained at 2, then after 3 months it decreases to 1 for a reverse doubling time of 1. Therefore this PSA doubling measure was appropriate for our patient.

(B) PSA isoforms. Example includes free PSA (fPSA, from which we can calculate the ratio of free PSA over bound PSA: %fPSA); proPSA, intact PSA (iPSA), and benign PSA (bPSA) [7–10]. Recent studies support the utility of %fPSA to reduce the number of biopsies in men with PSA levels between 2 and 10 ng/mL [11]. The risk of having prostate cancer is around 56% when %fPSA is below 10% but only 8% when it is over 25% [11]. Hence biopsies would be highly recommended in those men with <10% fPSA, but at a total PSA between 2 and 10 ng/mL [12]. Because our patient has already a PSA of 26.2 ng/mL at the time of diagnosis, this ratio was not performed and biopsies were already indicated. Therefore none of the PSA isoforms mentioned above were useful for our patient but could be useful for other patients.

Comment #1: Benign PSA appears to be a marker for benign prostatic hyperplasia (BPH) [11,13], but the clinical utility for prostate cancer diagnosis is not demonstrated. Studies on intact PSA are still very limited; one study found a correlation between lower proportions of intact PSA (iPSA) and a more advanced cancer stage, which could suggest a potential role as a marker for cancer aggressiveness [14]. The best-known usage of iPSA at the moment is as one of the four components of the 4Kscore test discussed below [8].

However, as mentioned above the patient PSA level was already at 26.2 ng/mL (not in the range of 2.0–10 ng/mL), as it was the first time our patient came to the walk-in clinic with symptoms of prostate cancer (see Case Report #1). Therefore it was too late for him to use these PSA isoforms to help making diagnosis of prostate cancer.

(C) Prostate health index (PHI). The PHI is simply the following mathematical formula: $([-2]proPSA/fPSA) \times \sqrt{PSA}$. It combines all three based PSA biomarkers into a single score. The rationale behind the formula is that men with higher levels $[-2]proPSA$ and PSA as well as with lower levels of fPSA are more likely to have prostate cancer [15].

Recent studies provide evidence that PHI substantially improves the screening capabilities of its individual counterparts [5,15,16]. In addition, PHI score seems to have excellent reliability across populations. In a large multicentric study, Catalona et al. observed increasing detection rates with increasing PHI scores with no effect of age or race, suggesting its applicability to all men irrespective of age and ethnicity [17]. Furthermore, several studies have observed a correlation between PHI scores and Gleason score on biopsies [15,17,18]. These findings support the use of PHI scores to reduce the number of unnecessary biopsies and as a measure of cancer aggressiveness during active surveillance [5,8].

One study of prostate cancer surveillance using PHI has demonstrated some evidence as a model predicting which men would be reclassified to higher-risk disease on repeat biopsies using baseline and longitudinal PHI scores (C-indices: 0.788 and 0.820 respectively at a median follow-up of 4.3 years) [19]. Similar results had been obtained by Isharwal et al. a year prior, using baseline PHI to predict unfavorable biopsy finding (C-index of 0.691) [20]. Today the PHI blood test is fairly inexpensive, simple, and is currently approved by the FDA, the EMA, and Health Canada and recommended by the National Comprehensive Cancer Network (NCCN) Guidelines for Prostate Cancer Early Detection [6,12].

The clinical utility is optimal for those in the diagnostic "gray zone" of 4–10 ng/mL PSA. PHI scores below 26 are associated with a 10% probability of cancer; approximately 17% risk for those in the 26–36 range; 33% risk for the 36–55 range and the probability of cancer for scores above 55 increases to above 50% [15,17].

Therefore PHI blood tests following a PSA score in the gray zone range (2–10 ng/mL) could help avoid overtreatment of low-risk prostate cancer. These would not only reduce unnecessary harm caused to low-risk patients, but also reduce the overall cost of prostate cancer detection while improving the quality of life of these patients [8].

Further studies are necessary to improve specificity of the score ranges as well as to clarify the accuracy of PHI in active surveillance. As mentioned above, our patient presented with a PSA level cannot be considered in the "gray zone" on his first visit to the walking clinic, suggesting that it was already too late for him to perform this score and so this measure becomes irrelevant. Again, this test does not apply to our patient, because his first biopsy was already positive for a highly undifferentiated and invasive prostate cancer, as explained above. However, this information remains useful for other patients with prostate cancer.

(D) 4Kscore. The 4Kscore (4KS) is another blood test looking to discriminate between indolent and aggressive prostate cancer by yielding an important risk of a high-grade cancer (Gleason score ≥ 7) on the following biopsies. It differs from PHI as it incorporates total PSA, free PSA, intact PSA, and the kallikrein-related peptide hK2 into the equation. Moreover, 4KS considers clinical information such as the age, prior negative biopsy status and negative digital rectal examination to provide a percent risk score [21]. The rationale for using these four kallikreins proteins as cancer detection and active surveillance tools comes from the evidence that their level increases with cancer cell undifferentiating and, either directly or indirectly, indicate prostate cancer progression and metastasis [22,23]. However, the 4Kscore is not yet FDA or Health Canada approved. Regardless, this biomarker was not really useful for our patient because at the time of presentation, the prostate biopsies were already positive for a Gleason score of 9, and the patient had already lung and bone metastasis.

(E) PC antigen 3 and Progensa. The Progensa prostate cancer antigen 3 (PCA3) assay is a nucleic acid amplification test that measures the urine concentration of (PCA3) and PSA RNA molecules, following a digital rectal examination (DRE). The ratio of PCA3 to PSA RNA yields the PCA3 score [6,22–24]. A PCA3 score below 25 is associated with a decrease in likelihood of prostate cancer, but decreasing the cutoff to a score of 10 reduced false positives by a little more than a third while false negatives increased only by 5.6%. Two different reviews summarizing 11 clinical studies determined that the overall accuracy of PCA3 was around 66% [8,25,26]. These promising results combined with easy specimen collection following DRE make it a very interesting biomarker. The assay is also part of the European Association of Urology guidelines for repeat biopsy decision making [27]. However, recent analysis indicates that PCA3 to be inferior to other biomarkers such as PHI and 4KS for malignant prostate cancer detection as in our patient. It seems that the most optimal gains of Progensa were obtained when used in combination with other biomarkers like TMPRSS2:ERG or within a multivariable model [25–29].

Despite being available commercially and approved by regulatory agencies, PCA3 is not commonly used as a first-line test in clinical practices and was not used in our patient. Again, our patient had already a high PSA level, a high Gleason score, and metastasis at the time of diagnosis, which renders this test useless for him. Gen-Probe received Health Canada licensure for the PROGENSA PCA3 assay based on a prospective, multicenter clinical study of the assay that enrolled 507 men. Gen-Probe submitted a Medical Device License Application to Health Canada in December of 2010. In the United States, the Progensa assay has been approved by the FDA since 2012. **(F) C-reactive protein (CRP).** CRP is an acute-phase protein released primarily by hepatocytes and is involved in the processes of inflammation, necrosis, and carcinogenesis. It has been associated with a poor prognosis of survival in several types of cancers [30–32]. Recent studies have looked into its potential predictive role in assessing prostate cancer severity and survival outcomes. Several of these studies, including a metaanalysis, have concluded that elevated CRP is associated with poor survival in prostate cancer patients with a cut-off value around 8.6 mg/L [30–32]. A single study of 261 patients has found a decrease in cancer-specific survival for those with elevated CRP (hazard ratio (HR) = 3.34) while the metaanalysis pooling 9 studies with >1400 patients obtained a HR of 1.91 for the same conclusion [30–32].

It is important to mention, however, that given the known association of CRP with several diseases related to survival (e.g., cardiovascular and pulmonary diseases), the above results should be interpreted with caution. Not all of those studies assessed potential comorbidities that could confound the relationship of high levels of CRP and low survival. In addition, different types of treatment (e.g., radiotherapy, chemotherapy, etc.) can trigger differential intensity of inflammation and could also play a role in the increase CRP levels observed [30]. Despite those limitations, it seems that for whatever reason a CRP count above 8.6 mg/L represents bad news for prostate cancer patients and should be interpreted accordingly.

For our patient, this biomarker was very useful and has been measured from the beginning of his treatment with hormone therapy. It was at 73.6 mg/L at his first visit to his primary care provider. Considering that our patient was not known for any cardiovascular, pulmonary, or other inflammatory diseases, his high CRP level was therefore predictive of a severe prostate cancer disease.

In fact, from the beginning, our patient was already having a grade IV prostate cancer with a Gleason score of 9. Because of this value or the cancer itself, the patient was prescribed Celebrex (Celecoxib; an NSAID) that gradually decrease the CRP level to normal. Finally, by taking this antiinflammatory drug, the CRP level becomes useless for the follow-up of cancer severity in our patient.

Question #2: Among the following, which are the most relevant biomarkers for the follow-up treatment of our patient's advanced prostate cancer?

(A) Circulating tumor cells
(B) Neutrophil-to-lymphocyte ratio (NLR)
(C) Circulating testosterone levels
(D) Androgen receptor splice variant 7
(E) All of the above

Answer: E.

(A) Circulating tumor cells (CTCs). CTCs are cells that have detached from a primary tumor and are circulating within the lymphatic or systemic blood vessels. These CTCs have been associated with an increased risk of metastases [33–35]. A study from de Bono and his group has determined that patients with metastatic castrate-resistant prostate cancer with a CTC count ≥5 (within a 7.5 mL sample of blood) had a much worse overall survival than those with lower CTC count. In addition the survival predictive value of CTC surpassed that of monitoring a decrease in PSA level [33–35].

One of the most interesting potential applications of CTCs as biomarkers being studied is its use as a treatment response indicator [8]. A recent study analyzed the treatment outcome of 486 patients from two major studies, all with CTC ≥ 5, and concluded that a ≥30% decline in CTC was associated with increased survival compared to those with a stable or increased CTC [34]. The cut-off value of 5 has been questioned recently by several studies that suggest that it would be best to simply interpret CTC as a continuous variable with a greater number representing a worse prognosis at all time points. Indeed, it seems that the relationship between the CTC count and survival is inversely proportional, regardless of the cutoff value chosen [35]. Unfortunately, this test was not performed for our patient, but it could have been useful for him and for other patients with advanced prostate cancer.

(B) Neutrophil-to-lymphocyte ratio (NLR). Similarly to CTCs the tumor microenvironment, the mix of local cells and immune cells around the tumor, is known to influence on cancer progression and treatment outcomes. Recently the NLR has gained interest from researchers for its prognostic value specifically for castration-resistant prostate cancer. First two independent studies, one with CYP17 inhibitor ketoconazole and the other with docetaxel chemotherapy, stratified their patients using a cut-off NLR value of 3. In both cases, pretreatment NLR >3 patients had a worse progression-free survival and overall survival, respectively, compared to patients with NLR $\leq$ 3 [36,37]. The interest generated by those findings led to the recent publishing of a metaanalysis on the matter. By compiling 22 studies, Cao et al. observed that a high NLR predicts a lower PSA response, a higher risk of recurrence and a worse overall survival, progression-free survival and recurrence-free survival in both metastatic castration-resistant prostate cancer and nonmetastatic sensitive prostate cancer [38].

Even though the metaanalysis revealed that there is still no clearly defined NLR cut-off value; most studies used NLR ˃ 3 or ˃ 5. The body of evidence to date supports the use of NLR value for risk stratification and personalization of treatment interventions for all types of prostate cancer patients. A patient with CTC and NLR scores above 5 might benefit from a more aggressive treatment regimen early on to maximize its survival, whereas a patient with low scores and no metastasis as yet could decide to withhold treatment for a while and improve his/her quality of life [8].

For our patient, this NLR score was already at 2.3 from the beginning and was suggestive of a good progression-free and overall survival. In December 2017, his NLR score goes down to 1.2, which suggests that that the first-line hormone therapy was an appropriate treatment, even though our patient becomes less responsive to hormone therapy. Therefore this biomarker can be used to follow the response to hormone therapy and is useful in indicating whether a more aggressive treatment regimen should be provided to our patient.

(C) Circulating testosterone levels. Along with surgery, chemotherapy and radiation therapy, hormonal therapy is the first-line treatment for locally advanced prostate cancer. Prostate tumor cells are particularly responsive to androgens (e.g., testosterone and dihydrotestosterone). As such, several drugs lowering testosterone levels have been approved by Health Canada for the hormone therapy of prostate cancer [2]. Circulating testosterone levels would then be an obvious choice of biomarkers to monitor the success and efficacy of those therapies, as well as to predict the chances of recurrence.

However, the association between serum testosterone and prostate cancer growth is rather weak. Some studies have found a correlation between increased cancer growth and higher levels of serum free testosterone [39], but several independent groups have found contradicting evidence [40,41]. Indeed, in a study of 168 patients, lower testosterone levels were associated with a higher Gleason score, which suggests the opposite association [40,42].

Our patient noticed that his testosterone level was adequately suppressed from the beginning, suggesting at some point that the hormone therapy he was taking was able to adequately suppress the testosterone. However, his testosterone level is currently still very low, but our patient is responding less well to hormone therapy. This confirms that other androgenic hormones might be involved in prostate cancer. Therefore going for a second-line hormone therapy such as Zytiga might be sufficient to reestablish the treatment response to hormone therapy by inhibiting similar androgenic hormones but at different levels (see Case Report #6).

(D) Androgen receptor splice variant 7. On the brighter side a promising new biomarker, AR-V7 (androgen receptor splice variant 7), seems an excellent biomarker to predict the response to antiandrogen treatments. In an important study of 31 patients treated with abiraterone acetate (Zytiga, a CYP17A1 inhibitor antiandrogen) and 31 patients treated with enzalutamide (Xtandi, an androgen receptor antagonist), CTCs were analyzed for presence of AR-V7. The biomarker was detected in 19% of abiraterone acetate patients and 39% of enzalutamide patients. Of all patients with AR-V7 positive receptors, none had a decrease in PSA levels and had a significantly lower overall survival compared with those having AR-V7 negative receptors [43].

The same group pursued their analysis and observed no difference with taxane chemotherapy in patients having the AR-V7 receptor positives compared to those being AR-V7 receptor negatives [43]. Hence they concluded that CTCs analysis for presence AR-V7 could be a first step in personalization of treatment options where AR-V7 positive patients would be started on taxane chemotherapy instead of antiandrogen drugs. Knowing the presence of this biomarker from the beginning may have influenced the choice for the initial treatment. For instance, if our patient is AR-V7 positive, it indicates that taxane should have been introduced earlier in his treatment, but this marker was not tested in our patient.

This recent research also further supports the importance of evaluating AR-V7 in our attempt to move toward an even more evidence-based personalized medicine. As our patient is now becoming resistant to hormone therapy and he is currently treated with abiraterone acetate, this biomarker becomes highly relevant for him.

Discussion: This case report reviewed the evidence concerning prostate cancer biomarkers currently available and with the greatest potential clinical utility in three distinct areas: detection and diagnosis; treatment and prognosis and metastasis surveillance and detection. However, it is by no means an exhaustive list of all the potential prostate cancer biomarkers in the literature; included here are those considered most useful for our patient. Several other biomarkers not discussed in this article are also very promising, and some will be briefly presented below.

The fusion gene TMPRSS2:ERG is promising, especially for its predictive value for both tumor progression and response to treatment [8]. Earlier we discussed the immune response regulator PD-1 (programmed cell-death receptor 1) as another biomarker gathering research attention. In a study performed in 2017 that included 535 prostate cancer patients a high density of PD-1+ lymphocytes was associated with worse survival outcomes, especially in patients with high initial Gleason scores >9 [44]. The list is growing and there is no doubt that as our understanding of immunotherapy grows, so too will be our ability to detect and use relevant biomarkers [2].

As highlighted by this article, several biomarkers that are already used in clinical practice seem to provide benefits for patients and for the health care system. Unfortunately the ever-increasing number of possible tests to perform gradually creates a difficult decisional puzzle for clinicians when it comes time to choose which diagnostic test to perform or which biomarkers to evaluate [8].

Establishing the superiority of a biomarker over another or demonstrating their additive prognosis power could make the decisional algorithms much simpler for health care providers, save money in the health care system by decreasing the necessity for multiple tests, and improve substantially the quality of life of prostate cancer patients via an improved individualized therapy [5,8]. The future of cancer therapy and cancer biomarkers is optimistic, but that future could come faster by investing more resources into consolidating the knowledge we already possess [5,8].

Health Canada has developed guidelines describing recommendations regarding the context, structure, and format of regulatory submissions for qualification of biomarkers (see ICH-E16 at the Resources for the Readers section). A biomarker qualification application might be submitted to regulatory authorities if the biomarker directly or indirectly helps in regulatory decision making. The objective of these guidelines is to create a harmonized recommended structure for biomarker qualification applications that will foster a consistency of applications across regions and facilitate discussions with and among regulatory authorities. It will also reduce the burden on sponsors, as a harmonized format will be recommended for use across all ICH regulatory regions [5,8].

It is also expected that the proposed document format will facilitate the incorporation of biomarker data into specific product-related applications. Biomarker qualification can take place at any time during chemical drugs or biotechnology derived product development, ranging from discovery through postapproval.

Multiple new immunotherapies, including vaccines and immune checkpoint inhibitors, are currently under investigation, and the demand for predictive and surrogate biomarkers will most certainly increase in the forthcoming years. Such biomarkers could identify responders in the earlier phases of treatments, in which the full effects are often not apparent for weeks or months after initiation. Because OS benefits are generally better demonstrated with immunotherapy than PFS benefits, such biomarkers could also provide surrogate endpoints to trials that would otherwise take years to complete [8].

REFERENCES

[1] Mondo DM, Roelh KA, Loeb S, et al. Which is the most important risk factor for prostate cancer: race, family history, or baseline PSA level? J Urol 2008;179:148.
[2] American Cancer Society. What's new in prostate cancer research? Available at: https://www.cancer.org/cancer/prostate-cancer/about/new-research.html; 2017.
[3] Moyer VA, Force USPST. Screening for prostate cancer: U.S. Preventive Services Task Force recommendation statement. Ann Intern Med 2012;157:120–34.
[4] Bell N, Connor Gorber S, Shane A, et al. Canadian Task Force on Preventive Health Care. Recommendations on screening for prostate cancer with the prostate-specific antigen test (guidelines). CMAJ 2014;186:1225–35.
[5] Archambault W, Plourde G. Biomarkers in advanced prostate cancer. J Pharmacol Clin Res 2017;3:1–9.

[6] Froehner M, Buck LM, Koch R, Hakenberg OW, Wirth MP. Derivatives of prostate-specific antigen as predictors of incidental prostate cancer. BJU Int 2009;104:25–8.

[7] Ayyildiz SN, Ayyildiz A. PSA, PSA derivatives, proPSA and prostate health index in the diagnosis of prostate cancer. Turk J Urol 2014;40:82–8.

[8] Gaudreau PO, Stagg J, Soulières D, Saad F. The present and future of biomarkers in prostate cancer: proteomics, genomics, and immunology advancements. Supplementary issue: biomarkers and their essential role in the development of personalised therapies (A). Biomark Cancer 2016;8(Suppl. 2):15–33.

[9] Makarov DV, Loeb S, Getzenberg RH, Partin AW. Biomarkers for prostate cancer. Annu Rev Med 2009;60:139–51.

[10] Roddam AW, Duffy MJ, Hamdy FC, Ward AM, Patnick J, Price CP, et al. Use of prostate-specific antigen (PSA) isoforms for the detection of prostate cancer in men with a PSA level of 2-10 ng/ml: systematic review and meta-analysis. Eur Urol 2005;48:386–99 [discussion 98–9].

[11] Carter HB, Partin AW, Luderer AA, Metter EJ, Landis P, Chan DW, et al. Percentage of free prostate-specific antigen in sera predicts aggressiveness of prostate cancer a decade before diagnosis. Urology 1997;49:379–84.

[12] McDonald ML, Parsons JK. The case for tailored prostate cancer screening: an NCCN perspective. JNCCN 2015;13:1576–83.

[13] Slawin KM, Shariat S, Canto E. BPSA: a novel serum marker for benign prostatic hyperplasia. Rev Urol 2005;7(Suppl. 8):S52–6.

[14] Peltola MT, Niemela P, Vaisanen V, Viitanen T, Alanen K, Nurmi M, et al. Intact and internally cleaved free prostate-specific antigen in patients with prostate cancer with different pathologic stages and grades. Urology 2011;77:1009.e1–8.

[15] Loeb S, Sanda MG, Broyles DL, Shin SS, Bangma CH, Wei JT, et al. The prostate health index selectively identifies clinically significant prostate cancer. J Urol 2015;193:1163–9.

[16] Hori S, Blanchet JS, McLoughlin J. From prostate-specific antigen (PSA) to precursor PSA (proPSA) isoforms: a review of the emerging role of proPSAs in the detection and management of early prostate cancer. BJU Int 2013;112:717–28.

[17] Catalona WJ, Partin AW, Sanda MG, Wei JT, Klee GG, Bangma CH, et al. A multicenter study of [−2]pro-prostate specific antigen combined with prostate specific antigen and free prostate specific antigen for prostate cancer detection in the 2.0 to 10.0 ng/ml prostate specific antigen range. J Urol 2011;185:1650–5.

[18] Lazzeri M, Haese A, Abrate A, de la Taille A, Redorta JP, McNicholas T, et al. Clinical performance of serum prostate-specific antigen isoform [−2]proPSA (p2PSA) and its derivatives, %p2PSA and the prostate health index (PHI), in men with a family history of prostate cancer: results from a multicentre European study, the PROMEtheuS project. BJU Int 2013;112(3):313–21.

[19] Tosoian JJ, Loeb S, Feng Z, Isharwal S, Landis P, Elliot DJ, et al. Association of [−2] proPSA with biopsy reclassification during active surveillance for prostate cancer. J Urol 2012;188:1131–6.

[20] Isharwal S, Makarov DV, Sokoll LJ, Landis P, Marlow C, Epstein JI, et al. ProPSA and diagnostic biopsy tissue DNA content combination improves accuracy to predict need for prostate cancer treatment among men enrolled in an active surveillance program. Urology 2011;77:763.e1–6.

[21] Punnen S, Pavan N, Parekh DJ. Finding the wolf in sheep's clothing: The 4Kscore is a novel blood test that can accurately identify the risk of aggressive prostate Cancer. Rev Urol 2015;17:3–13.

[22] Steuber T, Nurmikko P, Haese A, Pettersson K, Graefen M, Hammerer P, et al. Discrimination of benign from malignant prostatic disease by selective measurements of single chain, intact free prostate specific antigen. J Urol 2002;168:1917–22.

[23] Denmeade SR, Sokoll L, Dalrymple S, Rosen DM, Gady AM, Bruzek D, Ricklis RM, Isaacs JT. Dissociation between androgen responsiveness for malignant growth vs. expression of prostate specific differentiation markers PSA, hK2, and PSMA in human prostate cancer models. Prostate Cancer Prostatic Dis 2003;54:249–57.

[24] Sartori DA, Chan DW. Biomarkers in prostate cancer: what's new? Curr Opin Oncol 2014;26:259–64.

[25] Luo Y, Gou X, Huang P, Mou C. The PCA3 test for guiding repeat biopsy of prostate cancer and its cut-off score: a systematic review and meta-analysis. Asian J Androl 2014;16(3):487–92.

[26] Vlaeminck-Guillem V, Ruffion A, Andre J, Devonec M, Paparel P. Urinary prostate cancer 3 test: toward the age of reason? Urology 2010;75(2):447–53.

[27] Mottet N, Bellmunt J, Briers E, et al. Diagnostic evaluation: European Association of Urology. Available from: http://uroweb.org/guideline/prostate-cancer/#5; 2017. Accessed July 21, 2017.

[28] Tosoian JJ, Patel HD, Mamawala M, Landis P, Wolf S, et al. Longitudinal assessment of urinary PCA3 for predicting prostate cancer grade reclassification in favorable-risk men during active surveillance. Prostate Cancer Prostatic Dis 2017;20:339–42.

[29] Sanda MG, Feng Z, Howard DH, Tomlins SA, Sokoll LJ, Chan DW, et al. Association between combined TMPRSS2:ERG and PCA3 RNA urinary testing and detection of aggressive prostate cancer. JAMA Oncol 2017;3(8):1085–93.

[30] Liu ZQ, Chu L, Fang JM, Zhang X, Zhao HX, Chen YJ, et al. Prognostic role of C-reactive protein in prostate cancer: a systematic review and meta-analysis. Asian J Androl 2014;16:467–71.

[31] Thurner EM, Krenn-Pilko S, Langsenlehner U, Stojakovic T, Pichler M, Gerger A, et al. The elevated C-reactive protein level is associated with poor prognosis in prostate cancer patients treated with radiotherapy. Eur J Cancer 2015;51:610–9.

[32] Hall WA, Lawton CA, Jani AB, Pollack A, Feng FY. Biomarkers of outcome in patients with localized prostate cancer treated with radiotherapy. Semin Radiat Oncol 2017;27:11–20.

[33] de Bono JS, Scher HI, Montgomery RB, Parker C, Miller MC, Tissing H, et al. Circulating tumor cells predict survival benefit from treatment in metastatic castration-resistant prostate cancer. Clin Cancer Res 2008;14:6302–9.

[34] Lorente D, Olmos D, Mateo J, Bianchini D, Seed G, Fleisher M, et al. Decline in circulating tumor cell count and treatment outcome in advanced prostate cancer. Eur Urol 2016;70:985–92.

[35] Krebs MG, Hou JM, Ward TH, Blackhall FH, Dive C. Circulating tumour cells: their utility in cancer management and predicting outcomes. Ther Adv Med Oncol 2010;2:351–65.

[36] Keizman D, Gottfried M, Ish-Shalom M, Maimon N, Peer A, Neumann A, et al. Pretreatment neutrophil-to-lymphocyte ratio in metastatic castration-resistant prostate cancer patients treated with ketoconazole: association with outcome and predictive nomogram. Oncologist 2012;17(12):1508–14.

[37] Templeton AJ, Pezaro C, Omlin A, McNamara MG, Leibowitz-Amit R, Vera-Badillo FE, et al. Simple prognostic score for metastatic castration-resistant prostate cancer with incorporation of neutrophil-to-lymphocyte ratio. Cancer 2014;120(21):3346–52.

[38] Cao J, Zhu X, Zhao X, Li XF, Xu R. Neutrophil-to-lymphocyte ratio predicts PSA response and prognosis in prostate cancer: a systematic review and meta-analysis. PLoS ONE 2016;11(7):e0158770.

[39] Parsons JK, Carter HB, Platz EA, Wright EJ, Landis P, Metter EJ. Serum testosterone and the risk of prostate cancer: potential implications for testosterone therapy. Cancer Epidemiol Biomark Prev 2005;14:2257–60.

[40] Morgentaler A. Testosterone and prostate cancer: an historical perspective on a modern myth. Eur Urol 2006;50:935–9.

[41] Stattin P, Lumme S, Tenkanen L, Alfthan H, Jellum E, Hallmans G, et al. High levels of circulating testosterone are not associated with increased prostate cancer risk: a pooled prospective study. Int J Cancer 2004;108:418–24.

[42] Schatzl G, Madersbacher S, Thurridl T, Waldmuller J, Kramer G, Haitel A, et al. High-grade prostate cancer is associated with low serum testosterone levels. Prostate 2001;47:52–8.

[43] Antonarakis ES, Lu C, Wang H, Luber B, Nakazawa M, Roeser JC, et al. AR-V7 and resistance to enzalutamide and abiraterone in prostate cancer. N Engl J Med 2014;371:1028–38.

[44] Ness N, Andersen S, Khanehkenari MR, Nordbakken CV, Valkov A, Paulsen EE, et al. The prognostic role of immune checkpoint markers programmed cell death protein 1 (PD-1) and programmed death ligand 1 (PD-L1) in a large, multicenter prostate cancer cohort. Oncotarget 2017;8:26789–801.

CHAPTER 10

Case Report #10—Euthanasia (Assisted Suicide) or Palliative Sedation

INTRODUCTION

Our patient visits the clinic complaining of persistent vertigo and headaches, continual bone pain, and new urinary symptoms that are progressively getting worse for the last 4 years or longer. More than the pain, it is the reduction of his mobility and his ability to perform physical activities that he was used to do, though his mental activities have remained acceptable. He is concerned about the fact that he is seeing his body deteriorate without the possibility of improvement; combined with skeletal pains and psychological pain, this constitutes an unbearable suffering for him. Although the patient is not in a depressed mood, he does not want to be a burden for his family and the community when he will no longer be able to take care of himself adequately, mainly during the last months of his life. He also does not want to be oversedated and not conscious of what is happening around him while he suffers. As a physician, he is used to taking care of other people, but having people take care of him seems difficult to accept, and he does not want others to pity him. Furthermore, his family and friends are far away and he does not expect many visits, as they also have their family to care for, in addition to work and other commitments. Therefore the patient has questions about euthanasia and assisted suicide. Although the patient is far from being ready for this treatment, he wants to know his options in advance and he is afraid of not having access easily to such a treatment when he is ready to die.

Question #1: What are the main elements the patient should consider in this decision?

(A) The right to die is not entirely a personal decision
(B) The right to die in dignity is possible

Prostate Cancer
https://doi.org/10.1016/B978-0-12-815966-8.00010-2

© 2018 Elsevier Inc.
All rights reserved.

(C) **Does not want to become a burden for his family and for the society**

(D) **He does not want to die hooked up to a bunch of machines**

(E) **Having the right to die gives him the control, he needs to die in peace**

(F) **Not obliging to endure uncontrollable pain**

(G) **Good palliative care does not include the option of euthanasia**

(H) **Palliative sedation is highly acceptable**

(I) **All of the above**

Answer: I.

(A) The right to die is not entirely a personal decision. As stated by opponents of euthanasia the right to die obviously involves third parties: primary care providers, clinical team, parents, and friends who are involved in the care of a patient in his last days of life. It is not only the medical community that will agree to do it; there is a great number of pros and cons to the decision. Given the rising costs of the health care system, patients may think that they should ask for euthanasia to reduce the financial pressure on the health care system, especially if they consider that they are no longer useful to the community. The suffering patients might also consider that they are becoming more and more a burden for their family. This will be re-discussed below in Item C.

Opponents also consider that being too open to assisted suicide will put in danger the lives of several others, including people with chronic and progressive disabilities, degenerative diseases, or any other serious illness that may find this a viable option. This is based on the concern that euthanasia could lead to significant unintended changes in our health care system and society at large that we would later come to regret. These opponents also consider that asking doctors, nurses, or any other health care professional to carry out acts of euthanasia or assisted suicide could be a violation of their fundamental medical ethics. Finally, they consider that there is no reason a person should suffer either mentally or physically, as there are other effective end-of-life treatments available (palliative sedation), so euthanasia is not a valid treatment option for them and instead represents a failure on the part of the health care professionals and health care system involved in this patient's care.

Our patient is not totally in accordance with these concerns, as he is questioning why euthanasia or assisted suicide cannot be offered earlier; why does he have to tolerate uncontrollable physical and psychological pains to get access to euthanasia? When the physical and psychological pain become uncontrollable, patients should have the right to request euthanasia or assisted suicide earlier when this approach may represent the best treatment option for him/her instead of continuing to suffer without possibilities of health improvements.

The law provides very specific conditions according to which a person can receive euthanasia or assisted suicide. For instance, the patient should be of age and able to consent to care. He also should be at the end of his life and be suffering from a severe and incurable condition with no possibility of improvement. His medical situation should be characterized by advanced and irreversible capacities, and he should be suffering from constant and unbearable physical or psychological pains that cannot be addressed medically.

Unfortunately, when the patient is lacking support or when the support becomes exhausted, it is imperative that this patient should have access to euthanasia earlier. Furthermore, I am against considering the spouse or a son/daughter as criminals if they help a highly suffering parent or friend to die. This is also a sign of compassion rather than a criminal act, and the justice system should also demonstrate compassion for these people in such suffering situations. I think the government, public health agencies, and other National health agencies should provide more information and education to the public about euthanasia.

(B) The right to die in dignity. There is nothing particularly worthy to receiving an injection in order to induce a quick death, which is not always without pain, according to the opponents. Euthanasia in itself does not always save dignity; it induce the death of a person who often suffer from loneliness and the feeling of being a burden to the family and the community, according to the opponents of euthanasia.

On the other hand, dignity does not depend on the patient's physical or mental health, autonomy, or his usefulness to the society. Human dignity is based on the inherent worth of each human person, a value that must not be influenced by other circumstances or external factors such as money, work, and others. The simple fact of being human confers a dignity and the right to choose to continue to live with pains or to die.

One should consider that palliative care and palliative sedation also offer death with dignity because they provide patients with relief from pain as well as the social and emotional support patients need in order to face their death with courage. This will be discussed again below at items G and H.

This support requires, of course, time and perseverance. We are made of relationships capable of loving and caring for others. The feeling that we have maintained our dignity is linked to the respect we have for each other as humans. If people have the impression that they are losing their dignity, it is for the community to make sure they feel appreciated again. We all have the power to respond to the illness of others in friendship, love, and solidarity in order to support and protect their "right to life" until their natural death occurred. Obviously, we need each other in death as in life.

(C) No need to become a burden for my family or for society. This way of reasoning suggests that those who suffer are not worth the time and care they require, and our patient has often this impression in many circumstances. We must approach people with debilitating illness with compassion and not from a perspective of utility. We have the responsibility to love and support each other to ensure that no one feels obliged to request euthanasia or assisted suicide because he feels he is a burden on society or harbors feelings of loneliness. The fear of becoming a useless weight for family and society is the main reason mentioned by patients when they feel they are being asked to hasten their death. Many patients feel abandoned and are very isolated, but they should be comforted, encouraged, and supported.

(D) He does not want to die hooked up to a bunch of machines. In Canada the law does not require anyone to accept treatment. A patient who is considered apt to make his own decision or the agent of a patient legally allowed to make decisions for this patient has the right to accept or refuse any treatment. If treatment is interrupted or the decision is made not to administer treatment, then the cause of death is secondary to the disease progression of the patient and has nothing to do with euthanasia or assisted suicide. There is a big difference between letting someone die and inducing his death.

Artificial feeding and hydration are considered regular care, not a treatment, and must in principle be provided to patients. Indeed, water and food are essential for life and are not used to treat a specific condition. No one should die for having been deprived of food or water. In certain circumstances, however, as in the end of life the body may not be able to absorb the water and food. In this context, we can suspend artificial feeding and hydration.

It is true that the discontinuation of artificial nutrition and hydration at the end of life raises many questions among the general public and in the health care community. It may also make family members quite uncomfortable. Combined with the use of palliative sedation, it can raise questions of an ethical nature. However, having a clear understanding of and being able to clearly explain nutrition and hydration needs at the end of life to family members greatly reduce these concerns. Note that artificial nutrition and hydration are legally also considered to be treatments to which the patient can consent or refuse.

The natural progression of a serious disease is generally accompanied in the final days of deterioration in the patient's overall condition along with weakness, increasing bed confinement, loss of appetite, weight loss, dysphagia, and dyspnea. The patient often refuses to eat, and drinking becomes difficult. In these situations, appropriate mouth care prevents discomfort from a dry mouth. At this point, while systematic reviews of medical literature have not conclusively shown that nutrition and hydration are harmful for a person at the end of life, clinical experience seems to demonstrate that they can cause the patient additional discomfort.

This can lead to:

- increased peritumoral edema and, consequently, secondary pain
- increased edema, effusion, and ascites
- increased salivary, bronchial, and gastrointestinal secretions, thus increasing the incidence of terminal rales, nausea, and vomiting.

In fact, continuous palliative sedation rarely involves withholding nutrition or hydration; instead, they are usually discontinued spontaneously by the patient. To learn more about palliative sedation at the end of life, visit: http://www.cspcp.ca/wp-content/uploads/2017/11/Quebec-guidelines.pdf.

(E) Having the right to die gives him the control; he needs to die in peace. Allowing patients to die in peace is not control, but acceptance. It is important to provide people who suffer with the compassion and assistance they need in order to move toward acceptance until the time of death.

Requests for euthanasia and assisted suicide are often born of a deep sense of despair; it is usually a cry for help. As mentioned above the basis of these requests is the fear of pain or being alone in suffering.

Dying patients who are no longer able to decide for themselves could be seen as the attending physicians and their family members are taking control and decide for them. This could occur, for example, if someone had made a living will to clearly indicate his desire to be euthanized in certain circumstances, but then changes his mind when actually faced with these circumstances.

The alleged right to choose death could give others the right to impose your former decision once a patient can no longer show that he has changed his mind. Although this might be true if the living will has been clearly discussed from the beginning and documented in the patient's medical record, there is a minor possibility for this situation to occur. In this current case report, our patient has mandated two individuals to decide for him in this situation; a member of his family and a good friend not known by the family. In this situation the chances of nonrespect of the living will are further reduced.

(F) Not forcing to endure uncontrollable pain. Uncontrollable pain is very rare; it is usually possible to relieve most pain (see below under palliative sedation). However, as discussed above, euthanasia for patients with uncontrollable physical and psychological pain is also a sign of compassion, and the possibility to have access to assisted suicide should be made available earlier in some circumstances.

In the last moments of life, when the dying person's suffering is difficult to control (e.g., acute or chronic severe pain, dyspnea, nausea/vomiting, agitation, anxiety, and others), many medical organizations urge health care providers to provide immediate relief to the patient. It is unacceptable to allow a patient to suffer in the last days of his life. In this situation, it may be decided by the medical team, the patient, and the family to induce euthanasia or assisted suicide. This can be performed, for instance, by giving midazolam (benzodiazepine), then lidocaine (local anesthetic) or magnesium sulfate (anticonvulsant) in combination with propofol (anesthetic) or phenobarbital (barbituric), and then a neuromuscular blocker such as cisatracurium to make the patient died. This is just an example of a combination the attending physician can provide with assisted suicide; there are other combinations available. It is not the purpose of this book to discuss this issue, as the combination of drugs is decided by the medical teams taking care of a specific patient; the combination of drugs is most often established on a case by case basis.

According to this clinical practice guideline, the physician's interventions must always be directed at providing the best possible relief in situations of severe suffering that require analgesics, anxiolytics, or neuroleptics. The attending health care providers must not hesitate to rapidly consult a colleague who is an expert in end-of-life symptom relief if the treatments given to the patient fail to produce the desired response. Also, health care providers must not withhold relief from a dying patient for fear of causing secondary sedation when all other therapeutic options have failed.

(G) Good palliative care does not include the option of euthanasia. Euthanasia or assisted suicide is incompatible with the principles and goals of palliative care. Patients who enter the hospital hoping to receive adequate treatment for their last months, weeks, or days of life should not have to worry about being euthanized if it is not his/her request.

Incorporating euthanasia into palliative care is misleading the public about the true nature of palliative care. Palliative care is designed to ensure the best quality of life to the patient suffering from an incurable illness until death occurs.

Good palliative care can help the dying patient find a meaning to their pain in their suffering moments. The last weeks and days of life are often a time of reconciliation with family and friends. By choosing to interrupt a life prematurely, it may prevent the person from being exposed to profound human experiences, joy, and peace To learn more about palliative sedation at the end of life, visit http://www.cspcp.ca/wp-content/uploads/2017/11/Quebec-guidelines.pdf.

(H) Palliative sedation is highly acceptable. "Palliative sedation" is defined as the use of sedative medications to relieve refractory symptoms by a reduction in consciousness. Depending on the level of consciousness, three levels of sedation may be distinguished: mild, intermediate, and profound. Depending on the duration, sedation is often described as intermittent or continuous. Used adequately, analgesics rarely shorten life. The patient usually dies as a result of disease progression but without uncontrollable suffering.

Indeed the primary goal of palliative sedation is to alleviate patient suffering, regardless of its nature. Yet it is clear that at the end of life, clinical situations of extreme suffering or distress are happening that are difficult to relieve despite adequate palliative care. Some symptoms may be impossible to control, despite the extensive therapeutic resources used. They can worsen as death approaches, compromising the possibility of a peaceful death and adding to the distress of family members. Symptoms may be so severe that communication becomes impossible when the patient is experiencing such suffering. In these circumstances, medically induced sedation may be the most appropriate intervention. In other words, sometimes the only way to ensure comfort is to put the patient to sleep using pharmacological means; this is different from euthanasia. For more information, please consult the following link at: http://www.cmq.org/page/en/Soins-medicaux-fin-de-vie.aspx.

Commonly used agents include benzodiazepines, anticholinergics, neuroleptics (antipsychotics), barbiturates, and propofol. The pharmacological agents used to provide relief are administered safely, and the doses are adjusted in proportion to the patient's needs. The depth of sedation needed is determined based on the symptom in question and the level of suffering of the patient. This is determined on a case-by-case basis and can vary over time depending on the patient's response.

As the overriding goal of continuous palliative sedation in a patient nearing death is his comfort the parameters observed are mainly comfort oriented. Measurements of blood pressure, temperature, or oxygen saturation, which do not contribute directly to the patient's comfort, should be suspended. Observing the respiratory rate, however, is helpful to ensure the absence of tachypnea that could lead to respiratory distress. In addition, sudden-onset respiratory depression attributable to palliative sedation should be monitored and the dose adjusted, as this side effect is not desirable. For more information, please consult the following link: http://www.cspcp.ca/wp-content/uploads/2017/11/Quebec-guidelines.pdf.

During palliative sedation, other patient care techniques are maintained and contribute to the patient's comfort, such as patient mobilization, monitoring and rotation of subcutaneous or intravenous sites, monitoring for pressure ulcers, wound dressing, mouth care, verification of bladder emptying and intestinal transit, monitoring of infusion bags and tubing, monitoring for rales or signs of discomfort, and others. However, care that is no longer necessary must be discontinued. The patient must be carefully reassessed on a regular basis because he is no longer able to complain about his discomfort. For more information, please consult the following link: http://www.cspcp.ca/wp-content/uploads/2017/11/Quebec-guidelines.pdf.

Question #2: The patient is also questioning the use of cannabis for the control of his pain. As there are plenty of painkillers available on the market, he is not convinced that the use of cannabis is useful or appropriate for him. For which of the following conditions may cannabis be used in palliative care?

(A) Chemotherapy-induced nausea and vomiting (CINV)
(B) Cancer-associated pain
(C) Anorexia and cachexia syndrome
(D) Insomnia
(E) Depression and anxiety
(F) Cannabis as an antineoplastic agent
(G) All of the above
(H) Only A to E

Answer: H.

(A) Chemotherapy-induced nausea and vomiting (CINV). The endocannabinoid system is composed of receptors that are present within the emetic reflex pathways, making them a promising target for managing CINV. According to Turgeman and Bar-Sela, the central regulation of emesis occurs via the dorsal vagal complex (DVC), which includes three main areas: the area postrema; the nucleus of the solitary tract (nTS); and the dorsal motor nucleus of the vagus [1]. Located just outside the blood–brain barrier in the fourth ventricle of the brain the area postrema provides direct communication between blood-borne signals such as chemotherapy and the autonomic neurons that elicit emesis [1].

According to Turgeman and Bar-Sela, all three regions of the DVC, as well as its vagal outputs in the gastrointestinal tract, are populated with cannaboid (CB)-1 receptors, which have shown antiemetic effects when activated by the delta-9-tetrahydrocannabinol (Δ9-THC) [1].

Interestingly the (Δ9-THC) has shown opposite effects on the nTS compared to nausea-inducing 5-HT3 agonists [2]. In a metaanalysis by Machado et al. [3], they have observed in cancer patients receiving chemotherapy that dronabinol (Δ9-THC) has superior antiemetic activity than neuroleptics, while other cannaboid-based medicines (CBMs) had a clinical but not statistical advantage. However, CBMs have shown greater activity in suppressing anticipatory nausea than 5-HT3 antagonists [3].

Data from studies investigating the synthetically available versions of Δ9-THC have provided more convincing evidence. A quantitative systematic review [4] that included 30 randomized comparisons of oral nabilone, oral dronabinol, or the intramuscular levonantradol preparation (no longer available) with a placebo in 1366 patients receiving chemotherapy found that as antiemetics, Δ9-THC were more effective than prochlorperazine, metoclopramide, chlorpromazine, thiethylperazine, haloperidol, domperidone, or alizapride with a risk ratio of 1.38 (95% confidence interval: 1.18–1.51) [4].

However, adverse effects were noted as more intense and occurring more frequently in patients using cannabinoids. A recent systematic review [5] of 28 randomized controlled trials using nabilone or dronabinol and involving 1772 participants reported an overall benefit for cannabis. A Cochrane review [6] analyzed 23 randomized controlled trials of cannabinoids compared with a placebo or with other antiemetic drugs. Patients were more likely to report a complete absence of nausea and vomiting with cannabis than with a placebo, and there was little discernable difference between the effectiveness of cannabinoids and of prochlorperazine, metoclopramide, domperidone, and chlorpromazine. However, none of the trials involved the agents now most widely used: the serotonin 5-HT$_3$ antagonists [1].

The National Comprehensive Cancer Network guidelines cautiously mention cannabinoids as a breakthrough treatment for CINV patients who are not responsive to CINV or to other antiemetics [7]. This suggests that cannabis could be a good alternative for our patient with advanced prostate cancer at the time he will receive chemotherapy, if that occurs.

Comment #1: Cannabinoids (CBDs) may be extracted naturally from the plant and taken in herbal form or manufactured synthetically. They can be inhaled, smoked, or injected, as well as mixed with food or tea. Cannabinoids exert its mental and physical effects by binding to G protein-coupled cannabinoid receptors 1 and 2 (CB1 and CB2) and then stimulating the endogenous cannabinoid system and altering the levels of endocannabinoids (eCBs) [1]. These receptors are widely distributed throughout the body, with the highest concentration of CB1 and CB2 found in the central nervous system and in the immune cells, respectively [1].

According to Bar-Sela et al., CBs are neuroactive lipid messengers that contribute to physiological processes such as rewards, motivation, memory, learning, and pain [8]. This is an interesting topic for our patient, an obesity expert, as he has already been involved in the research and development of a CB1 receptor antagonist for the treatment of obesity, which has been found to be able to decrease appetite and regulate body weight in humans. Therefore A is a good response.

(B) Cancer-associated pain. According to Turgeman and Bar-Sela, cannabis has long been used for its analgesic properties [1]. The CB-1Rs is largely present in the hippocampus area and its associated cortices, the cerebellum, and the basal ganglia, and they have similar neuroanatomy, neurochemical, and pharmacological characteristics to the receptors of the opioid system [1]. They are thought to modulate nociceptive processing in the brain, both independently and in synergism with exogenous opioids. The CB-2Rs located in the dorsal root ganglion of sensory neurons in the spinal cord may also have a role [1]. They stimulate the release of analgesic beta endorphins and reduce C-fiber (i.e., the fiber responsible for identifying pain and transmitting that information to the central nervous system) activity in neuropathic pain models [1].

According to Fine and Rosenfeld, peripheral cannabinoid receptors have been implicated in antinociception by activating noradrenergic pathways [9]. Research made by Noyes et al., suggested that cannabis is a potent therapeutic adjunct, with some capacities for relieving neuropathic pain [10]. These authors demonstrated that high doses of Δ9-THC were significantly superior to placebo in pain reduction but comparable to codeine, although associated with considerable sedation. Several trials have examined the analgesic effects of Δ9-THC/CBD preparations in subjects with opioid refractory cancer pain. Among them, Portenoy et al. [11] found a higher proportion of patients who were reporting analgesia with low and medium dose nabiximols (e.g., Sativex; a specific extract of cannabis) than placebo, while poor drug tolerability was noted in the high-dose group [11].

Johnson et al. [12] have also observed superior pain relief in patients treated with Δ9-THC/CBD as compared to Δ9-THC alone or placebo, which was sustained for 2 years without the need for raising opioid dosages. Similarly, Bar-Sela et al. [13] performed a transversal observational study evaluating cancer-related symptoms reported by patients while on CBMs. They have found that not only was there less pain, but also a reduction in opioid dosage in close to half of the subjects, suggesting a better control of pain in patients with cancer [13].

Despite these positive results the standardization is difficult due to differing cannabis preparations and dosages; therefore larger trials are needed to delineate a more accurate picture [1]. Considering the data presented above, our patient may prefer to try cannabis instead of increasing doses of morphine if it is medically prescribed for the control of his pain, especially considering that bone metastasis is particularly painful in patients with prostate cancer.

Comment #1: As mentioned earlier the CB1 receptors that are initially identified in the brain are found in higher concentrations in areas involved in the processing of noxious stimuli. The CB2 receptors are predominantly located in cells of the immune system and likely have a role in the control of inflammation and cell proliferation [14]. It was suggested that the entire function of the system of cannabinoid receptors and endocannabinoids might be to assist in the modulation of the response to pain [15].

A recent systematic review consisting of 28 studies involving a total of 2454 participants was published, and the drug tested included inhaled cannabis, dronabinol, nabilone, and nabiximols, among others [16]. Twelve of the studies investigated neuropathic pain, and three looked at patients with cancer pain. The studies generally showed improvement in pain measures, with an overall odds ratio of 1.41 (95% confidence interval: 0.99–2.00) for improvement in pain with the use of cannabinoids as compared with a placebo [14].

An earlier systematic review of 18 randomized controlled trials of cannabinoids in 766 participants with chronic noncancer pain found that 15 of the studies reported a significant analgesic effect for the cannabinoids compared with a placebo, and a number of studies also noted improvements in sleep [14].

Another review that included 6 of those 18 studies in patients with cancer-related pain also favored cannabinoids [15]. Therefore cannabis represents a very good alternative to other opioids or in combination with other opioids for pain control. It also has the advantage of being associated with fewer side effects [15].

Comment #2: Neuropathic pain is often seen in cancer patients [16–18]. A systematic review of six randomized, double-blind, placebo-controlled trials of cannabinoids (five specifically addressing neuropathic pain) found evidence for the use of low-dose medical cannabis in refractory neuropathic pain in conjunction with traditional analgesics [16]. Another analysis reviewed five trials of inhaled cannabis in patients with HIV-related peripheral neuropathy and again found a positive effect for cannabis compared with a placebo [17]. A recent small study [18] showed a dose-response effect for vaporized cannabis in the relief of pain from diabetic peripheral neuropathy, which is a huge clinical problem estimated to affect 238 million people worldwide.

Comment #3: In animal models, cannabinoids and opioids have been known to have synergistic analgesic effects [19]. Analgesic effects of cannabinoids are not blocked by opioid antagonists, suggesting that the two types of agents work through different receptors and pathways, which suggest that the combination of opioids and cannabis could be a good approach for patients with cancer. An early study found that THC was ineffective as an analgesic on its own, but it increased slightly the effect of morphine, which is also a good approach for our patient [20]. A randomized controlled trial of dronabinol (a synthetic form of THC) in patients on opioids for chronic pain found that when compared with a placebo, dronabinol reduced pain ($P<.01$) and increased patient satisfaction ($P<.05$) [21]. A randomized controlled trial of nabiximols (a synthetic form of THC) in 359 cancer patients with poorly controlled pain despite a stable opioid treatment regimen found that the sublingual preparation (4, 10, or 16 sprays daily for 5 weeks) reduced both pain and sleep disruption [11].

Comment #4: Patients prescribed high doses of opiates at the end of life by their oncologist or palliative care team frequently feel totally unable to communicate with their loved ones in their precious remaining time because of altered cognition [1]. Many patients have successfully weaned themselves down or off of their opiate doses by adding cannabis to their treatment. Although it would seem that THC-dominant strains of cannabis would be most likely to have analgesic effects, patients often report significant pain reduction from strains that are predominantly CBD-rich [1]. Again, cannabis represents a good alternative for patients at the end of their life if palliative sedation is required, resulting in less drowsiness than that caused by opiate drugs. Therefore B is a good response.

(C) Anorexia and cachexia syndrome. Anorexia and cachexia in cancer patients refer to a spectrum of metabolic changes that begins with a reduced caloric intake and variable degrees of inflammation, then progresses to a refractory procatabolic state linked to a poor quality of life and short survival. In patients with cachexia, gaining body weight or even slowing the rate of weight loss is incredibly difficult [1]. Jatoi et al. [22] compared dronabinol and megestrol acetate (progestatif) for cancer-associated anorexia, with significant positive findings in favor of megastrol. In fact a trial comparing Δ9-THC to Δ9-THC + CBD to placebo found no significant improvements in survival, weight, or other nutritional variables [23]. However, cannabis has been associated with improved taste, smell, and food enjoyment. But he might need cannabis in the future to improve taste and food enjoyment.

The cachexia syndrome comes with weight loss, muscle loss, fatigue, weakness, and appetite loss [5]. As a result, treating both the wasting syndrome and the underlying cause are of main importance. We should note that standard treatment for cachexia is not very effective in itself, as this condition is usually secondary to another illness. Therefore the treatment of cachexia usually involves treating the underlying cause [5].

However, clinicians recommend a metabolite called HMB (beta-hydroxy beta-methylbutyrate) for reducing the loss of muscle mass in patients with cancer [5]. Additionally, as discussed in a previous case report, health care providers recommend a high protein diet or protein supplements for patients suffering from cancer. Beyond that, primary care providers use appetite stimulants to encourage patients to eat more in order to increase their body weight and muscle mass. Many health care professionals in various countries suggest that cannabis helps with this syndrome. A study shows that Dutch health care professionals of various specialties prescribe the use of medicinal cannabis for an array of ailments [1]. One of these is anorexia and cachexia, which is associated with cancer. The study also notes that smoked marijuana was perceived as being more effective than oral administration. Regardless the cannabis was used with the aim of stimulating appetite and/or increasing body weight [5].

Another study notes that dronabinol, a CBM, and cannabis cigarettes have shown numerous benefits in the treatment of wasting syndrome in HIV patients. They note that cannabis was significantly superior to the placebo for increasing or maintaining body weight [5]. Over the course of the study, those patients taking dronabinol did not lose any weight; the placebo group, on the other hand, did. They add that cannabinoids, the active chemical compounds found in cannabis, were effective in treating both a lack of appetite and weight loss in those suffering from wasting syndrome.

By increasing appetite and, more importantly, body weight, cannabis can improve the quality of life of patients and slow or halt the deterioration of some ailments.

One small study of dronabinol in cancer patients demonstrated an enhanced chemosensory perception in the treatment group compared with the placebo group [23]. In the dronabinol recipients, food tasted better, and appetite and caloric intake increased. Therefore cannabis should be considered for our patient with advanced prostate cancer. The fact that cannabis is promoting better food taste and appetite is certainly relevant, especially considering that at the time of his stage IV prostate cancer diagnosis, our patient lost >15 pounds in a few months and had difficulty eating because of his perceived poor taste of food. Therefore C is a good response.

(D) Insomnia. A large metaanalysis by Whiting et al. [5] has reviewed 19 studies that evaluated sleep as an outcome as well as two trials specifically investigating sleep problems. The authors found a positive association between cannabinoids and improved sleep quality. Therefore our patient may opt for this treatment considering that he his having problems with insomnia. However, meditating for 20 minutes or more before going to bed improves his sleep.

According to Project CBD, some patients with sleep issues report that "ingesting a CBD-rich tincture or extract a few hours before bedtime has a positive effect that facilitates a good night of sleep." The key is finding the right strain, blend, product, and dose for each individual patient. Everyone responds to cannabis differently, so it may take a little trial and error before finding the perfect fit. Therefore D is a good response.

For more information about Project CBD, please consult the following link: https://www.projectcbd.org/about/about-project-cbd.

(E) Depression and anxiety. The metaanalysis performed by Whiting et al. [5] where depression was not included in this metaanalysis in cancer patients found no difference compared to the placebo, with the exception of a negative effect for high dose nabiximols in one. Considering that our patient is having chronic bone pain due to the metastases causing at some points some symptoms of depression and anxiety, cannabis represents a potential therapeutic choice for him to help fight episodes of depression and anxiety, as the time for action is faster than with antidepressants that may take weeks before becoming fully effective. Therefore CBD is a good option for treating depression and anxiety [5].

(F) Cannabis as an antineoplastic agent. The anticancer potential of cannabis has been explored in preclinical research with evidence of a connection to cancer cell-signaling pathways. It has been discussed in the review made by Turgeman and Bar-Sela that cannabis can induce apoptosis (tumor cell death) and inhibit tumor proliferation, tumor vascularization, and metastasis [1]. In fact, in nonsmall-cell lung carcinoma for instance, the administration of Δ9-THC inhibited endothelial growth factor-induced migration in vitro, as well as tumor and metastasis growth in mice models [24]. Secondly, in murine gliomas, CBD had antiproliferative effects, while selective CB2 agonists caused tumor regression [24]. Thirdly the activation of CB1 in mice with colon carcinoma reduced tumor growth. Finally, treatment with cannabinoids has also been associated with reduced tumor growth in models of breast cancer, hepatocellular carcinoma, and multiple hematological malignancies [24]. However, research on cannabis as an antineoplastic medication is still lacking. Considering this lack of information on this potential beneficial effect of cannabis in the treatment of cancer, cannabis does not represent a therapeutic option, at this time, for our patient or for other patients suffering from cancer.

In addition to the well-known palliative effects of cannabinoids on some cancer-associated symptoms, a huge body of evidence shows that these molecules can decrease tumor growth in animal models of cancer [24]. They do so by modulating key cell-signaling pathways involved in the control of cancer cell proliferation and survival. In addition, cannabinoids inhibit angiogenesis and decrease metastasis in various tumor types in laboratory animals [24].

Discussion: This case report has briefly discussed the main components associated with physical and psychological pain controls and a peaceful death. The purpose of this case report is also to help our patient and other patients in making the appropriate decision when facing an end-of-life situation. The aspects related to medication to induce palliative sedation or euthanasia have only been succinctly presented here, as these treatments are performed on a case-by-case basis and with an informed decision between the patient and the treatment physicians or medical teams. Following discussions with the attending health care providers, the patient, and family, the acceptance of these treatment modalities should result from a positive and informed benefit/risk assessment.

The information provided here illustrates a decision-making process that allows patients to make the right decision. To ensure an acceptable quality of life where euthanasia could be seen by the patients or their relatives, and by the health-care providers and other caregivers, as a necessary step to ensure quality of care, until the ultimate end occurred. To ensure that this decision process is adequately and legally respected by the attending physicians a specific form must be completed. This form is available via this link: http://www.cmq.org/pdf/outils-fin-de-vie/sedation-formulaire-declaration-eng.pdf?t=1473952844153.

When the patient has to choose between excruciating pain before dying and a faster death a faster death may actually appear as "quality of care" and as upholding the patient's dignity. Some physicians opposed to euthanasia are suggesting that primary care physicians provide terminal palliative sedation in order to allow the patient to die from a "natural" death. However, in the face of these principles, there are patients seeking an end to their suffering because there is no possible alternative. For them, fortunately, euthanasia (assisted suicide) or palliative sedation is available.

The concept of dignity, although universal in some ways, becomes very personal. This does not mean that the health care provider and relatives should not discuss with the patient, to see the beauty of life that remains. However, the ultimate assessment of quality of life should go back to the patient and his treating physician to decide whether to receive palliative sedation or euthanasia.

Studies have added support to the growing body of knowledge of cannabis use in palliative oncology care. However, these studies have many limitations. Promising data on pain, nausea, and vomiting relief, as well as a relatively favorable safety profile and potential anticancer properties, will allow for more focused research in the future.

For each medical decision patients have to make, they should perform a benefit/risk analysis. We have discussed the potential benefits above, but to have a complete picture regarding the use of cannabis as palliative therapy, we should be informed about the side effects associated with its use. Adverse effects of cannabis are mostly short term and include somnolence, dizziness, dry mouth, and disorientation, as well as euphoria, anxiety, and hallucinations. Memory and cognition problems, addiction, and exacerbation of psychiatric illness, such as depression and anxiety disorders, have also been associated with cannabis use [1,5]. Events are mostly attributed to Δ9-THC, while the opposing cannabinoid CBD is thought to alleviate these effects and instead facilitate learning, prevent psychosis, and ease anxiety. Unfortunately, street cannabis contains high levels of Δ9-THC and a negligible level of CBD, while CBD supplied for research or patient use have dramatically different potencies and cannabinoid proportions [1].

Most long-term effects of cannabis have been shown to subside within 6 weeks of abstinence from cannabis use [25]. That is probably the main reason that patients suffering from pain and anxiety should have access to the medically prescribed cannabis instead of getting it from the street or from friends. Furthermore, having the medically prescribed cannabis ensure a better benefit/risk ratio, since the data have been reviewed by Health Authorities which also include the contraindications of use, the specific warnings and precautions, the long-term effect of use, the considerations for specific populations and others safety measures that are not available for cannabis obtained from the street

In the meantime, clinical trials with cannabis are continuing in order to find the right drug composition, dosage, and means of administration per indication and patient. With evolving legislation, improved education and training, and increasing availability, medical cannabis will gradually be integrated into a standard, evidence-based oncology practice.

REFERENCES

[1] Turgeman I, Bar-Sela G. Cannabis use in palliative oncology: a review of the evidence for popular indications. IMAJ 2017;19:85–8.

[2] Himmi T, Dallaporta M, Perrin J, Orsini JC. Neuronal responses to delta 9-tetrahydrocannabinol in the solitary tract nucleus. Eur J Pharmacol 1996;312:273–9.

[3] Machado Rocha FC, Stéfano SC, De Cássia Haiek R, Rosa Oliveira LM, Da Silveira DX. Therapeutic use of *Cannabis sativa* on chemotherapy-induced nausea and vomiting among cancer patients: systematic review and meta-analysis. Eur J Cancer Care (Engl) 2008;17:431–43.

[4] Tramer MR, Carroll D, Campbell FA, Reynolds DJ, Moore RA, McQuay HJ. Cannabinoids for control of chemotherapy induced nausea and vomiting: quantitative systematic review. BMJ 2001;323:16–21.

[5] Whiting P, Wolff RF, Deshpande S, et al. Cannabinoids for medical use [review and meta-analysis]. JAMA 2015;313:2456–73.

[6] Smith LA, Azariah F, Lavender VT, Stoner NS, Bettiol S. Cannabinoids for nausea and vomiting in adults with cancer receiving chemotherapy. Cochrane Database Syst Rev 2015;11:CD009464.

[7] National Comprehensive Cancer Network (NCCN). NCCN Clinical Practice Guidelines in Oncology: Antiemesis. Fort Washington, PA: NCCN; 2017. Ver 2.2015. Available at: http://www.nccn.org/professionals/physician_gls/pdf/antiemesis.pdf.

[8] Bar-Sela G, Avisar A, Batash R, Schaffer M. Is the clinical use of cannabis by oncology patients advisable? Curr Med Chem 2014;21:1923–30.

[9] Fine PG, Rosenfeld MJ. The endocannabinoid system, cannabinoids, and pain [review]. Rambam Maimonides Med J 2013;4(4):e0022.

[10] Noyes Jr. R, Brunk SF, Avery DA, Canter AC. The analgesic properties of delta-9-tetrahydrocannabinol and codeine. Clin Pharmacol Ther 1975;18:84 89.

[11] Portenoy RK, Ganae-Motan ED, Allende S, et al. Nabiximols for opioid treated cancer patients with poorly-controlled chronic pain: a randomized, placebo controlled, graded-dose trial. J Pain 2012;13:438–49.

[12] Johnson JR, Lossignol D, Burnell-Nugent M, Fallon MT. An open-label extension study to investigate the long-term safety and tolerability of THC/CBD oromucosal spray and oromucosal THC spray in patients with terminal cancerrelated pain refractory to strong opioid analgesics. J Pain Symptom Manage 2013;46:207–18.

[13] Bar-Sela G, Vorobeichik M, Drawsheh S, Omer A, Goldberg V, Muller E. The medical necessity for medicinal cannabis: prospective, observational study evaluating treatment in cancer patients on supportive or palliative care. Evid Based Complement Alternat Med 2013;2013:510392.

[14] Lynch ME, Campbell F. Cannabinoids for treatment of chronic non-cancer pain: a systematic review of randomized trials. Br J Clin Pharmacol 2011;72:735–44.

[15] Martin-Sanchez E, Furukawa TA, Taylor J, Martin JL. Systematic review and meta-analysis of cannabis treatment for chronic pain. Pain Med 2009;10:1353–68.

[16] Deshpande A, Mailis-Gagnon A, Zoheiry N, Lakha SF. Efficacy and adverse effects of medical marijuana for chronic noncancer pain: systematic review or randomized controlled trials. Can Fam Physician 2015;61:e372–81.

[17] Andrea MH, Carter GM, Shaparin N, et al. Inhaled cannabis for chronic neuropathic pain: a meta-analysis of individual patient data. J Pain 2015;16:1221–32.

[18] Wallace MS, Marcotte TD, Umlauf A, Gouaux B, Atkinson JH. Efficacy of inhaled cannabis on painful diabetic neuropathy. J Pain 2015;16:616–27.

[19] Cichewicz DL. Synergistic interactions between cannabinoid and opioid analgesics. Life Sci 2004;74:1317–24.

[20] Naef M, Curatolo M, Petersen-Felix S, Arendt-Nielsen L, Zbinden A, Brenneisen R. The analgesic effect of oral delta-9-tetrahydrocannabinol (thc), morphine, and a thc-morphine combination in healthy subjects under experimental pain conditions. Pain 2003;105:79–88.

[21] Narang S, Gibson D, Wasan AD, et al. Efficacy of dronabinol as an adjuvant treatment for chronic pain patients on opioid therapy. J Pain 2008;9:254–64.

[22] Jatoi A, Windschitl HE, Loprinzi CL, et al. Dronabinol versus megestrol acetate versus combination therapy for cancer-associated anorexia: a North Central Cancer Treatment Group study. J Clin Oncol 2002;20:567–73.

[23] Brisbois TD, de Kock IH, Watanabe SM. Delta-9-tetrahydrocannabinol may palliate altered chemosensory perception in cancer patients: results of a randomized, double-blind, placebo-controlled pilot trial. Ann Oncol 2011;22:2086–93.

[24] Valesco G, Sánchez C, Guzmán M. Anticancer mechanisms of cannabinoids. Curr Oncol 2016;(Suppl. 2)S23–32.

[25] Curran HV, Freeman TP, Mokrysz C, Lewis DA, Morgan CJ, Parsons LH. Keep off the grass? Cannabis, cognition and addiction [review]. Nat Rev Neurosci 2016;17:293–306.

RESOURCES FOR THE READERS

Arthritis and musculoskeletal and skin disorders. 2016. Available at: https://www.niams.nih.gov/.

Canadian Alcohol and Drug Use Monitoring Survey (CADUMS). Available at: http://www.hc-sc.gc.ca/hc-ps/drugs-drogues/cadums-esccad-eng.php.

Canadian Center of Substance Abuse and Addictins. Available at: http://www.ccdus.ca/Pages/default.aspx.

Canadian Centre for the Fight Against Substance Abuse and Addiction. http://www.ccdus.ca/Pages/default.aspx.

Canadian Coalition for Action on Tobacco. Available at: http://www.plaintruth.ca/.

Canadian Food Guide. Available at: https://www.canada.ca/en/health-canada/services/canada-food-guides/revision-process.html.

Canadian Immunization Guideline. Available at: https://www.canada.ca/en/public-health/services/canadian-immunization-guide.html.

Canadian Society for Exercise Physiology (CSEP). Available at: http://www.csep.ca/home.

Canadian Task Force on Preventive Health Care. Available at: https://canadiantaskforce.ca/guidelines/published-guidelines/.

Continuous palliative sedation report form. Available at: http://www.cmq.org/pdf/outils-fin-de-vie/sedation-formulaire-declaration-eng.pdf?t=1473952844153.

Environmental and Workplace Health. Available at: http://www.hc-sc.gc.ca/ewh-semt/index-eng.php.

Health and Safety at Work Commission (Canadian Center for Occupational Health and Safety [CCOHS]). Available at: https://www.ccohs.ca/.

ICH-E16: Biomarkers related to drug or biotechnology product development: context, structure and format of qualification submissions. http://www.ich.org/products/guidelines/efficacy/article/efficacy-guidelines.html.

International Agency for Research on Cancer (WHO). Available at: https://www.iarc.fr/index.php.

Medical care on the last days of life. Available at: http://www.cmq.org/publications-pdf/p-1-2015-05-01-en-soins-medicaux-derniers-jours-de-la-vie.pdf.

Nationale Cancer Institute. Available at: https://www.cancer.gov/.

National Comprehensive Cancer Network guidelines. Available at: https://www.nccn.org/professionals/physician_gls/f_guidelines.asp.

National Institute of Arthritis and Musculoskeletal and Skin Disorders. https://www.niams.nih.gov/.

Nutrition Guide for Men with Prostate Cancer. Available at: http://www.bccancer.bc.ca/nutrition-site/Documents/Patient%20Education/nutrition_guide_for_men_with_prostate_cancer.pdf.

Occupational Cancer Research Center: Available at: http://www.occupationalcancer.ca/.

Palliative sedation at the end of life. Available at: http://www.cspcp.ca/wp-content/uploads/2017/11/Quebec-guidelines.pdf.

Procure Canada. Available at: http://www.procure.ca/en/.

Prostate Cancer Canada: Information, Testing, Treatment, Research ... Available at: http://www.prostatecancer.ca/.

Prostate Cancer Center of Excellence: Wake Forest School of Medicine. Available at: http://www.wakehealth.edu/Research/Comprehensive-Cancer-Center/Prostate-Cancer-Center-of-Excellence/Research.htm.

Special access program. Available at: https://www.canada.ca/en/health-canada/services/drugs-health-products/special-access/drugs.html.

Tobacco Free Initiative (TFI): the MPOWER measures. Available at: http://www.who.int/tobacco/mpower/package/en/.

United States Environmental Protection Agency. Available at: http://www.epa.gov/.

INDEX

Note: Page numbers followed by *f* indicate figures.

CPI Antony Rowe
Chippenham, UK
2018-12-04 17:15